Power Control Circuits Manual

Newnes Circuits Manual Series

Audio IC Circuits Manual   R.M. Marston
Diode, Transistor & FET Circuits Manual   R.M. Marston
Electronic Alarm Circuits Manual   R.M. Marston
Instrumentation & Test Gear Circuits Manual   R.M. Marston
Modern CMOS Circuits Manual   R.M. Marston
Modern TTL Circuits Manual   R.M. Marston
Op-amp & OTA Circuits Manual   R.M. Marston
Optoelectronics Circuits Manual   R.M. Marston
Power Control Circuits Manual   R.M. Marston
Timer/Generator Circuits Manual   R.M. Marston

Power Control Circuits Manual

2nd Edition

R. M. Marston

Newnes
An imprint of Butterworth-Heinemann

To Esther, the true power in the Marston household.

Newnes
An imprint of Butterworth-Heinemann
Linacre House, Jordan Hill, Oxford OX2 8DP
A division of Reed Educational and Professional Publishing Ltd

 A member of the Reed Elsevier plc group

OXFORD   BOSTON   JOHANNESBURG
MELBOURNE   NEW DELHI   SINGAPORE

First published 1990
Reprinted 1992, 1993, 1995
Second edition 1997

© R.M. Marston 1997

British Library Cataloguing in Publication Data
Marston, R. M. (Raymond Michael), 1937–
    Power control circuits manual – 2nd ed.
    1. Electronic circuits
    I. Title
    621.3'815

ISBN 0 7506 3005 1

Library of Congress Cataloguing in Publication Data
Marston, R. M.
    Power control circuits manual/R. M. Marston – 2nd ed.
        p.   cm. – (Newnes circuits manual series)
    Includes index.
    ISBN 0 7506 3005 1 (pbk.)
    1. Power electronics.   2. Electronic control.   I. Title.
    II. Series.
    TK7881.15 M37   96–42138
    621.31'7–dc20   CIP

Composition by Genesis Typesetting, Rochester, Kent
Printed and bound in Great Britain by
Biddles Ltd, Guildford and King's Lynn

Contents

Preface

Modern power control circuits can be used to adjust (either manually or automatically) the brilliance of lamps, the speed of motors, the temperature of electric heaters or radiators, or the loudness of audio signals, etc. This control can be achieved using electro-mechanical switches or relays, or electronic components such as transistors, SCRs, triacs, or power ICs, etc. This revised, updated and considerably enlarged second edition of the best-selling *Power Control Circuits Manual* takes an in-depth look at the whole subject of 'electronic' power control, and in the process presents the reader with 339 carefully selected circuits and diagrams, which are backed up by almost 50 000 words of concise, highly informative and very readable text.

The book is specifically aimed at the practical design engineer, technician, and experimenter, but will also be of great interest to the electronics student and amateur, and deals with its subject in an easy-to-read, down-to-earth, mainly non-mathematical but very comprehensive manner. Most of the book's chapters deal with some specific aspect of power control, and start off by explaining the basic principles of that aspect of the subject and then go on to present the reader with a wide range of practical application circuits.

The volume is split into nine distinct chapters. The first of these explains the basic principles of electrical/electronic power control, and the second shows practical control circuits using conventional switches and relays. Chapter 3 describes ways of using modern CMOS devices as low-power electronic switches, and Chapters 4 and 5 deal with AC and DC power control systems and present many practical application circuits. Chapter 6 describes ways of controlling DC motors, including those of the 'stepper' and servo types. Chapters 7 and 8 deal with audio power control and DC power supply systems, respectively. The final chapter presents a miscellaneous collection of AC power control data and circuitry, including basic information on modern house rewiring techniques.

Throughout the volume, great emphasis is placed on practical 'user' information and circuitry, and the book abounds with useful circuits and illustrations. In most AC power control circuits, alternative component values are (where applicable) given for use with 115V or 230V (nominal) AC supply lines. Most of the solid state devices used in the practical circuits are modestly priced and readily available types, with universally recognised type numbers.

R. M. Marston, 1996

—— 1 ——

Basic principles

An electrical or electronic power control circuit can be defined as any circuit that is used to control the distribution or the levels of AC or DC power sources. Such circuits can be used to manually or automatically control the brilliance of lamps, the speed of motors, the temperature of electric fires, heaters or soldering irons, or the loudness of audio signals, etc., or they can be used to manually switch power to these or other devices, or to switch power automatically when parameters such as temperature or light intensity, for example, go beyond pre-set limits.

A variety of devices can be used in power control applications. These range from simple switches and electromechanical devices such as relays and solenoids, which can be used as low-speed power switches, to solid-state devices such as transistors, FETs, CMOS multiplexers, SCRs or triacs, or power ICs, etc., which can be used as high-speed power switches or magnitude controllers. This opening chapter describes basic electronic power control principles, and shows how the above devices can be used in power control applications at levels ranging from a fraction of a milliwatt to several kilowatts.

Power switching circuits

All electric power controllers can be categorised as either power switchers (such as a lamp *on/off* switch) or power level controllers (such as a lamp dimmer). *Figure 1.1* shows examples of three basic types of power switching circuit, and *Figures 1.2* to *1.5* illustrate the operating principles of four different types of power level control circuit.

The three basic switching circuit types are the *on/off* controller (*Figure 1.1(a)*), which is used to switch power to a single load, the power distributor (*Figure 1.1(b)*), which switches power to one or other of several alternative loads, and the power selector (*Figure 1.1(c)*), which feeds one or other of several alternative power sources to a single load. Note in these circuits that power switching is shown via ordinary electric switches, but in practice these can easily be replaced by sets of relay contacts or by any of a variety of types of solid-state switch.

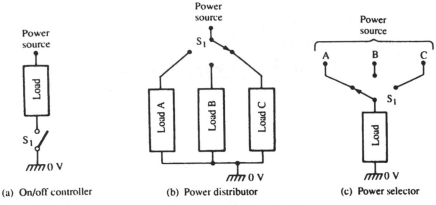

Figure 1.1   *Three basic types of power switching circuit*

DC power control

Figure 1.2 shows the basic circuit of a simple DC power level controller in which 0 to 12V is available on RV1 slider and is fed to the load via a current-boosting voltage follower buffer stage. This type of circuit is not very efficient, since all unwanted power is 'lost' across the buffer stage; if, for example, the load is fed with 1V and draws 1A (thus consuming 1W), 11V are lost in the 1A buffer, which thus consumes 11W, so the circuit operates with an efficiency of only 8.33 percent.

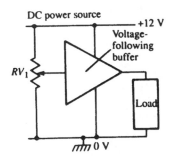

Figure 1.2   *Simple DC power level controller*

Figure 1.3 shows a different type of power level controller, which operates with an efficiency of about 95 percent. Here, power is fed to the load via a solid-state power switch that is activated via a squarewave generator with a variable mark–space (M/S) ratio or duty cycle. For explanatory purposes, assume that the duty cycle is variable from 5 to 95 percent via RV1, and that the solid-state switch is 100 percent efficient. In this case the circuit operates as follows.

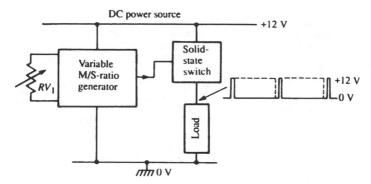

Figure 1.3 *Switched-mode DC power level controller*

When the solid-state switch is open, zero volts are generated across the load, and when it is closed the full 12V supply line voltage is generated across the load. When the switch is activated via the variable M/S-ratio generator, the *mean* voltage of the load (integrated over one duty cycle) is proportional to the generator's duty cycle. Thus, if the generator has a 50 percent duty cycle (i.e. a 1:1 M/S ratio, or equal ON and OFF times), the *mean* load voltage equals 50 percent of the 12V supply value, or 6V. Similarly, if the duty cycle is 5 percent, the mean load voltage is 600mV, and if the duty cycle is 95 percent the mean voltage is 11.4V. Since power consumption is proportional to the square of the mean supply voltage, it can be seen that this circuit enables load power to be varied from 0.25 to 90.25 percent of maximum via RV1.

In practice, a peak of only 200mV or so is usually lost across a solid-state power switch, so this type of circuit operates with a typical efficiency of about 95 percent at all times, and is widely used in DC lamp-brilliance and motor-speed control applications.

AC power control

Most domestic and small-business premises are supplied with single-phase AC power supplies (usually 240V, 50Hz in the UK, 115V, 60Hz in the USA), which are fed into the premises via a 3-core power cable. Some large businesses and many industrial units are also supplied with 3-phase 50Hz or 60Hz power supplies (for driving high-power electric motors, etc.), and most ships and aircraft have on-board AC generators that deliver 3-phase 400Hz to 800Hz outputs. Unless otherwise stated, all AC power control circuits shown in this volume are designed for use with standard single-phase 50Hz–60Hz power lines that provide r.m.s. output voltages in the range 110V to 250V.

Figures 1.4 and *1.5* show two ways of adapting the basic switched-mode variable-duty-cycle DC power control technique for use in single-phase AC power control applications. The *Figure 1.4* circuit uses a 'phase-triggered' switching technique

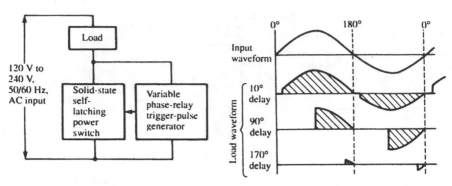

Figure 1.4   *Variable phase-delay-switching AC power controller, with waveforms*

widely used for controlling the AC power feed to filament lamps, etc., which have fairly long thermal time constants, and to electric power drills and motors, which have high mechanical inertia, and the *Figure 1.5* circuit uses a 'burst-fire' technique widely used for controlling electric fires, etc., which consume high currents and have long thermal time constants.

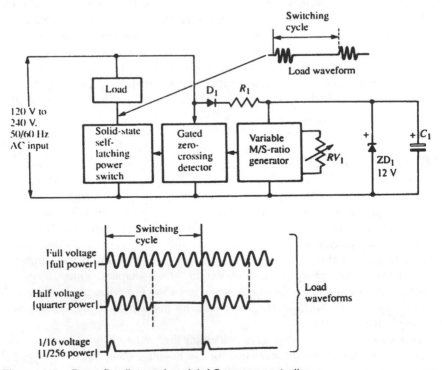

Figure 1.5   *Burst-fire (integral-cycle) AC power controller*

In *Figure 1.4*, power is applied to the load via a self-latching solid-state power switch (a triac, for example), which can be triggered (via a variable phase-delay network and a trigger pulse generator) at any point during each power half-cycle, but which automatically unlatches again at the end of each half-cycle as the AC voltage falls briefly to zero. The diagram shows the load voltage waveforms that can be generated.

Thus, if the power switch is triggered just after the start of each half-cycle (with near-zero phase delay), the mean load voltage equals almost the full supply value, and the load consumes near-maximum power. If the switch is triggered half way through each half-cycle (with 90 degrees phase delay), the mean load voltage equals half the supply value, and the load thus consumes one-quarter of maximum power. Finally, if the switch is triggered near the end of each half cycle (with near-180 degrees delay), the mean load voltage is almost zero, and the load consumes minimal power.

The *Figure 1.4* phase-triggered power control technique is highly efficient (typically, better than 95 percent), and enables the load power to be fully varied over a wide range; since switching occurs at the power line frequency, it enables lamp brilliance to be varied without flicker. Its major disadvantage is that, since power may be switched abruptly from zero to a high peak value (particularly at about 90 degrees delay), the resulting high current surges can generate substantial RFI (radio frequency interference); this type of circuit is thus not suitable for feeding high-current loads such as electric fires, etc.

Burst-fire control

High-current resistive loads such as electric heaters ('fires') can be efficiently power controlled without generating significant RFI by using the *Figure 1.5* burst-fire technique, in which power bursts of complete half-cycles are fed to the load at regular line frequency-related intervals. Thus, if bursts are repeated at 8-cycle intervals, the mean load voltage equals the full supply line value if the bursts are of 8-cycle duration, or half voltage (equals quarter power) at 4-cycle duration, or 1/16th voltage (equals 1/256th power) at one half-cycle duration, etc.

The burst-fire technique generates near-zero RFI because it switches power to the load only very near the start of line half-cycles, when the instantaneous line voltage (and load current) is very low. This is achieved by using a line-driven zero-crossing detector, which is gated via a variable M/S-ratio generator and gives an output only if it is gated on and the instantaneous line voltage is below 7V or so; the detector output triggers the self-latching solid-state device (triac) that switches power to the load. The variable M/S-ratio generator is powered from a 12V DC supply derived from the AC power line via D1-R1 and ZD1-C1.

The burst-fire or 'integral-cycle power control' technique is very efficient, but enables the load's power consumption to be varied only in a number of discrete half-cycle steps. When driving electric heaters, this last-mentioned factor is of little importance, and the system can easily be used to give precision automatic room-temperature control with the aid of suitable temperature-sensing thermistors or thermostats, etc.

Electric switch basics

The simplest type of power control device is the ordinary electric switch, which comes in several basic versions; a selection of these is depicted by the symbols of *Figure 1.6*. The simplest switch is the push-button type, in which a spring-loaded conductor can be moved so that it does or does not bridge (short) a pair of fixed contacts. These switches come in either normally-open (n.o. or NO) form (*Figure 1.6(a)*), in which the button is pressed to short the contacts, or in normally closed (n.c. or NC) form (*Figure 1.6(b)*), in which the button is pressed to open the contacts.

The most widely used switch is the moving arm type, which is shown in its simplest form in *Figure 1.6(c)* and has a single spring-loaded (biased) metal arm or 'pole' that has permanent electrical contact with terminal A but either has or has not got contact with terminal B, thus giving an ON/OFF switching action between these terminals. This type of switch is known as a single-pole single-throw (SPST) switch.

Figure 1.6(d) shows two SPST switches mounted in a single case with their poles 'ganged' together so that they move in unison, to make a double-pole single-throw (DPST) switch.

Figure 1.6(e) shows a single-pole double-throw (SPDT) switch in which the pole can be 'thrown' so that it connects terminal A to either terminal BA or BB, thus enabling the A terminal to be coupled in either of two different directions or 'ways'.

Figure 1.6(f) shows a ganged double-pole or DPDT version of the above switch. Note that these multi-way switches can be used in either simple ON/OFF or multi-way power distribution/selection applications.

Figure 1.6(g) shows a switch in which the A terminal can be coupled to any of four others, thus giving a '1-pole, 4-way' action. Finally, *Figure 1.6(h)* shows a ganged 2-pole version of the same switch. In practice, switches can be designed to give any desired number of poles and 'ways'.

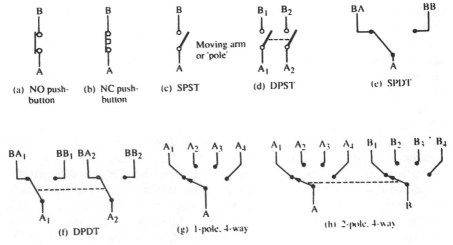

Figure 1.6   *Some basic switch configurations*

Two other widely used electric switches are the pressure-pad type, which takes the form of a thin pad easily hidden under a carpet or mat and which is activated by body weight, and the microswitch, which is a toggle switch activated via slight pressure on a button or lever on its side, thus enabling the switch to be activated by the action of opening or closing a door or window or moving a piece of machinery, etc.

Electromechanical relay basics

The conventional electromagnetic relay is really an electrically operated switch, and is a very useful power control device. *Figure 1.7* illustrates its operating principle. Here, a multi-turn coil is wound on an iron core, to form an electromagnet that can move an iron lever or armature which in turn can close or open one or more sets of

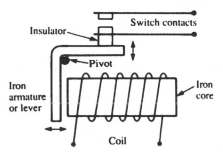

Figure 1.7   *Basic design of standard electromagnetic relay*

switch contacts. Thus, the operating coil and the switch contacts are electrically fully isolated from one another, and can be shown as separate elements in circuit diagrams.

The main characteristics of the relay coil are its operating voltage and resistance values, and *Figure 1.8* shows alternative ways of representing a 12V, 120R coil. The symbol of *Figure 1.8(c)* is the easiest to draw, and carries all vital information. Practical relays may have coils designed to operate from a mere few volts DC, up to the full AC power line voltage, etc.

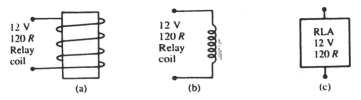

Figure 1.8   *Alternative ways of representing a 12 V, 120R relay coil*

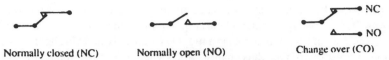

Figure 1.9 *The three basic types of contact arrangement*

There are three possible basic types of relay contact arrangement, these being normally closed (n.c. or NC), normally open (n.o. or NO), and changeover (c.o. or CO), as shown in *Figure 1.9*. Practical relays often carry more than one set of contacts, with all sets ganged. Thus, the term 'd.p.c.o. (or DPCO)' simple means that the relay carries two sets of changeover contacts. Actual contacts may have electrical rating up to several hundred volts, and up to tens of amps.

Relay configurations

Figures 1.10 to *1.13* show basic ways of using ordinary relays. In *Figure 1.10*, the relay is wired in the non-latching mode, in which push-button switch S1 is wired in series with the relay coil and its supply rails, and the relay closes only while S1 is closed.

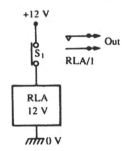

Figure 1.10 *Non-latching relay switch*

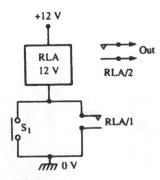

Figure 1.11 *Self-latching relay switch*

Figure 1.11 shows the relay wired to give self-latching operation. Here, n.o. relay contacts RLA/1 are wired in parallel with activating switch S1. RLA is normally off, but turns on as soon as S1 is closed, making contacts RLA/1 close and lock RLA ON even if S1 is subsequently reopened. Once the relay has locked on it can be turned off again by briefly breaking the supply connections to the relay coil.

Note in the above circuits that the relay can be switched in the AND mode via several series-wired switches, so that the relay turns on only when all switches are closed, or in the OR mode via several parallel-wired switches, so that the relay turns on when any of these switches are closed. *Figure 1.12* shows these modes used in a simple burglar alarm, in which the relay turns on and self-latches (via RLA/1) and activates an alarm bell (via RLA/2) when any of the S1 to S3 switches are briefly closed (by opening a door or window or treading on a mat, etc.). The alarm can be enabled or turned off via key switch S4.

Some relays coils can be activated via only a few volts and milliamps, enabling them to be turned on and off via simple transistor (or IC) circuitry if desired, as shown in the example of *Figure 1.13*, where a coil current of 100mA can be obtained via an S1 current of less than 4mA. Note that relay coils are highly inductive and can generate back-e.m.f.s of hundreds of volts if their coil currents are

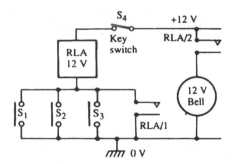

Figure 1.12 *Simple burglar alarm*

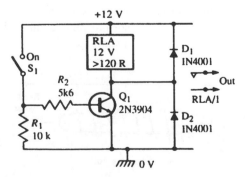

Figure 1.13 *Transistor-driven relay with two-diode coil damper*

suddenly broken; these back-e.m.f.s can easily damage electronic coil-driver circuitry. This danger can be overcome by connecting protective 'damping' diodes D1 and D2 to the coil as shown; D1 prevents the RLA-Q1 junction from swinging more that 600mV above the positive supply rail value, and D2 stops it from swinging more than 600mV below the zero-volt rail value.

Reed relay basics

Another type of electromechanical relay is the 'reed' type, which consists of a springy pair of opposite-polarity magnetic reeds with gold- or silver-plated contacts, sealed into a glass tube filled with protective gases, as shown in *Figure 1.14*. The opposing magnetic fields of the reeds normally hold their contacts apart, so they act as an n.o. switch, but these fields can be effectively cancelled or reversed (so that the switch closes) by placing the reeds within an externally generated magnetic

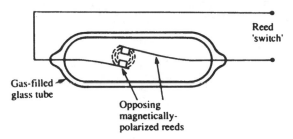

Figure 1.14   *Basic structure of reed relay*

field, which can be derived from either an electric coil that surrounds the glass tube, as shown in *Figure 1.15*, or by a permanent magnet placed within a few millimetres of the tube, as shown in *Figure 1.16*.

Practical reed relays are available in both n.o. and c.o. versions, and their contacts can usually handle maximum currents of only a few hundred milliamps. Coil-driven types can be used in the same way as normal relays, but typically have a drive-current sensitivity ten times better than a standard relay.

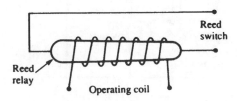

Figure 1.15   *Reed relay operated by coil*

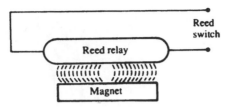

Figure 1.16 *Reed relay operated by magnet*

A major advantage of the reed relay is that it can be 'remote activated' at a range of several millimetres via an external magnet, enabling it to be used in many home-security applications; *Figure 1.17* illustrates the basic principle. The reed relay is embedded in a door or window frame, and the activating magnet is embedded adjacent to it in the actual door or window so that the relay changes state whenever the door/window is opened or closed: several of these relays can be interconnected and used to activate a suitable alarm circuit, if desired.

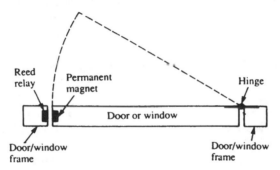

Figure 1.17 *Method of using a reed relay/magnet combination to give burglar protection to a door or window*

Transistor devices

So far only electromechanical power control devices have been described. Many solid-state devices can also be used in power control applications, and the simplest of these is the discrete bipolar transistor, which is usually used in the switching mode as shown in *Figures 1.18* and *1.19*. In the case of the npn transistor the switch load is wired between Q1 collector and supply positive, and in the case of the pnp device it is wired between Q1 collector and the zero-volts rail. In both cases the switch driving signal is applied to Q1 base via R1, which has a typical resistance about twenty times greater than the load resistance value.

In the npn circuit Q1 is cut off (acting like an open switch), with its output at the positive supply voltage value, when zero input signal applied, but can be driven to saturation (so that it acts like a closed switch and passes current from collector to

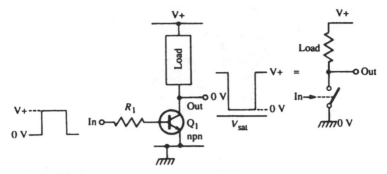

Figure 1.18 *Bipolar npn transistor switch circuit*

emitter) by applying a large positive input voltage, under which condition the output equals Q1's saturation voltage value (typically 200mV to 600mV).

The action of the pnp circuit (*Figure 1.19*) is the reverse of that described above, and Q1 is driven to saturation (with its output a few hundred mV below the supply voltage value) and passes current from emitter to collector when zero input drive voltage is applied, and is cut off (with its output at zero volts) when the input equals the positive supply rail value.

A useful variation of the transistor switch is the optocoupler, which comprises an infra-red LED (Light-Emitting Diode) and matching phototransistor, mounted close together (optically coupled) in a light-excluding package, as shown in the basic application circuit of *Figure 1.20*. Here, when SW1 is open zero current flows in the LED, so Q1 is in darkness and also passes zero current (it acts like an open switch), so zero output appears on R2. When SW1 is closed, current flows in the LED via R1, illuminating Q1 and making it act as a closed switch that generates an R2 output voltage which can thus be controlled via the R1 input current, even though R1 and R2 are fully isolated electrically. In practice, this device can be used to optocouple either digital (switching) or analogue signals, and offers hundreds of volts of isolation between the input and output circuits.

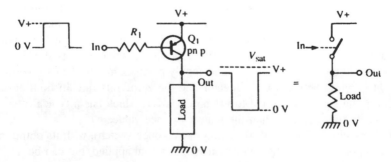

Figure 1.19 *Bipolar pnp transistor switch circuit*

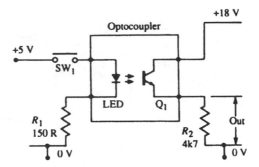

Figure 1.20 *Basic optocoupler switching circuit*

FET devices

A Field-Effect Transistor (FET) is 3-terminal voltage-controlled current generator with a near-infinite input or GATE impedance. FETs are available in two basic versions, giving either an enhancement or a depletion mode of operation. Enhancement mode FETs pass zero current when the gate voltage is zero, and the current rises (is enhanced) when the gate is forward biased. Depletion mode FETs give the reverse of this action; they pass a maximum current when the gate voltage is zero, and the current falls (depletes) when the gate is forward biased. Note in the rest of this book that only enhancement mode FETs are considered.

Practical FETs are available under several exotic names related to details of their construction; thus, names such as IGFET (Insulated Gate FET), MOSFET (Metal Oxide Silicon FET) and JFET (Junction FET) are widely used among low-power devices, and names like VMOS, HMOS and HEXFET are common among high-power devices.

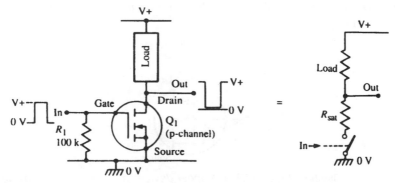

Figure 1.21 *Enhancement-mode n-channel MOSFET switching circuit*

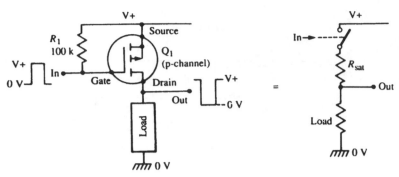

Figure 1.22 *Enhancement-mode p-channel MOSFET switching circuit*

FETs are available as n-channel and p-channel types, which are analogous to npn and pnp transistors respectively. *Figures 1.21* and *1.22* show how to use n-channel and p-channel enhancement-mode MOSFETs as simple electronic switches. The n-channel MOSFET acts as an open switch when its gate voltage is zero, and as a closed switch (in series with a 'saturation resistance') which passes current from *drain* to *source* when its gate is at the positive supply rail value. The p-channel MOSFET gives the reverse action, and acts as a closed switch (plus saturation resistance) which passes current from *source* to *drain* when its gate voltage is zero, and as an open switch when its gate is at the positive supply rail value.

The 'closed switch' saturation resistance of a MOSFET may vary from a few hundred ohms in a low-power device to a fraction of an ohm in a high-power one, and gives a voltage-divider action with the load resistor that determines the closed-state output voltage of the circuit. Thus, if the *Figure 1.21* circuit has a 10V supply and a 900R load, it will give a 'closed' output of 1V0 at an R_{SAT} value of 100R, or 10mV at an R_{SAT} value of 0R9.

Note that a MOSFET's gate terminal has a near-infinite input resistance, and if allowed to 'float' can accumulate electrostatic charges that can destroy the device. Some MOSFETS have a built-in zener diode to give protection against this danger, as shown in *Figure 1.23*.

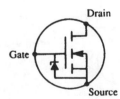

Figure 1.23 *Symbol of MOSFET n-channel device with internal zener diode gate protection*

Basic CMOS switch

One of the most important digital IC families is that known as CMOS, and this whole family is based on the simple Complementary MOSFET digital inverter circuit of *Figure 1.24*, which comprises nothing more than a p-channel and an n-channel enhancement-mode MOSFET wired in series between the two supply

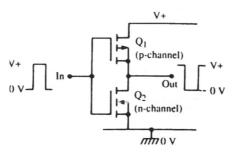

Figure 1.24 *Basic CMOS digital inverter circuit*

lines, with the two gates tied together at the input terminal and with the output taken from the junction of the two devices. In use, the input is at either zero volts (logic-0) or full positive supply rail voltage (logic-1).

Figure 1.25(a) shows the digital equivalent of this circuit with a logic-0 input, with Q1 acting as a closed switch in series with an R_{SAT} of 400R, and Q2 acting as an open switch; the circuit thus draws zero quiescent current but can 'source' fairly

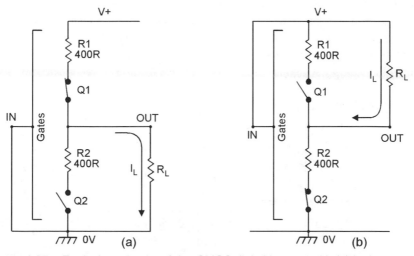

Figure 1.25 *Equivalent circuits of the CMOS digital inverter with (a) logic-0 and (b) logic-1 inputs*

large drive currents into an external output-to-ground load via the 400R output resistance (R1) of the inverter. *Figure 1.25(b)* shows the equivalent of the inverter circuit with a logic-1 input. In this case Q1 acts as an open switch, but Q2 acts as a closed switch in series with an R_{SAT} of 400R; this circuit thus draws zero quiescent current, but can 'sink' fairly large currents via an external supply-to-output load via its internal 400R output resistance (R2).

Thus, the basic CMOS digital inverter has a near-infinite input impedance, draws near-zero (typically 0.01μA) quiescent current with a logic-0 or logic-1 input, can source or sink substantial output currents, and has an output that is inherently short-circuit proof via its 400R output impedance. It gives an excellent low-power switching action.

CMOS bilateral switches

One very important member of the CMOS family is the bilateral switch or transmission gate, which is shown in basic and symbolic forms in *Figures 1.26(a)* and *1.26(b)*. The importance of this device is that (like a normal switch or set of relay contacts) it can conduct current in either direction (bilaterally), whereas a single transistor or FET can conduct in one direction only (from collector to emitter in the case of an npn device, or from emitter to collector in a pnp device).

The operation of the *Figure 1.26(a)* circuit is fairly simple. The device comprises an n-channel and a p-channel MOSFET wired in inverse parallel (drain-to-source and source-to-drain), but with their gate signals applied in anti-phase via a pair of CMOS inverter stages to give the bilateral switching action. Thus, when the control signal is at logic-0 the gate of Q2 is driven to logic-1 and that of Q1 is driven to logic-0, and under this condition both MOSFETs act as open switches between the circuit's X and Y points. When, on the other hand, the control signal is at logic-1 the gate of Q2 is at logic-0 and that of Q1 is at logic-1, so both MOSFETs act as closed switches, and a low resistance (equal to the R_{SAT} value) exists between the X and Y points of the circuit.

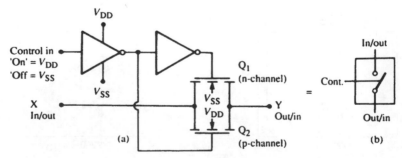

Figure 1.26   (a) *Basic circuit and* (b) *symbol of simple CMOS bilateral switch or transmission gate*

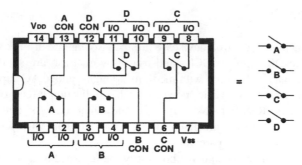

Figure 1.27 *The 4016B and 4066B quad bilateral switches each act as four independent SPST switches*

Note when the control input is at logic-1 that signal currents can flow in either direction between the X and Y terminals (via Q1 in one direction, or Q2 in the other), provided that the signals are within the logic-level voltage limits; the X and Y terminals can thus be used as either IN or OUT terminals.

Practical CMOS bilateral switch circuits are usually a bit more complex than shown in *Figure 1.26*, and give typical ON resistances of about 100R. All practical CMOS ICs of these types house several bilateral switches. The 4016B and 4066B ICs, for example, each house four such switches configured as independent SPST switches, as shown in *Figure 1.27*, and the 4051B houses a set of switches and logic network configured to act as a single-pole 8-way bilateral switch, as shown in *Figure 1.28*.

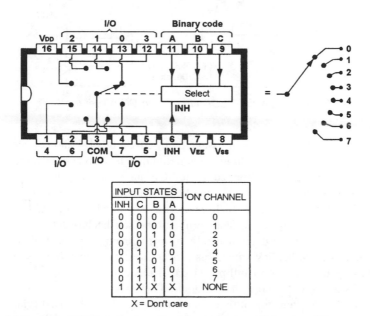

INPUT STATES				'ON' CHANNEL
INH	C	B	A	
0	0	0	0	0
0	0	0	1	1
0	0	1	0	2
0	0	1	1	3
0	1	0	0	4
0	1	0	1	5
0	1	1	0	6
0	1	1	1	7
1	X	X	X	NONE

X = Don't care

Figure 1.28 *The 4051B acts as a single-pole 8-way bilateral switch*

SCR basics

The solid-state power control devices examined so far are useful with low- to medium-voltage DC supplies. SCRs and triacs, by contrast, are solid-state power control devices intended for use with medium- to high-voltage AC supplies. *Figure 1.29* shows the symbol and basic 'power switch' application circuit of an SCR, or Silicon Controlled Rectifier. Note in this and other circuits that follow that alternative component values are given for use with 120V and 240V power lines, the 240V values being noted in parentheses.

The SCR can (as its name and symbol imply) be regarded as a 3-terminal silicon rectifier that can be controlled via its *gate* terminal. Normally, it acts as an open switch, but if its *anode* is positive to its *cathode* it can be made to switch on and act as a forward-biased rectifier by applying a brief trigger current to its gate terminal; if its resulting anode-to-cathode current exceeds a minimum 'holding' value (usually a few mA) the SCR self-latches into the 'on' state and stays there until its anode-to-cathode current falls below the minimum holding value, at which point the SCR reverts to the open-switch state. Thus, the *Figure 1.29* circuit acts as follows.

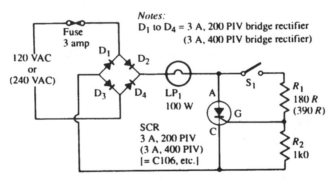

Figure 1.29   *Full-wave ON-OFF SCR circuit with DC power load*

The AC power line signal is full-wave rectified via D1–D4 and converted into a waveform that goes from zero to maximum and then back to zero again in each AC half-cycle; this waveform as applied to the SCR anode via lamp-load LP1. Thus, if S1 is open, zero gate drive is fed to the SCR, which acts like an open switch, and the lamp is off. If S1 is closed, SCR gate drive is applied via R1–R2, so just after the start of each half-cycle the SCR turns on and self-latches until the end of the half-cycle, at which point it automatically turns off again as its forward current falls below the minimum holding value; this process repeats in each half-cycle, and the lamp thus operates at almost full power under this condition.

The SCR anode falls to only a few hundred mV when the SCR turns on, so note that S1 and R1–R2 consume little mean power, but S1 can be used (via the SCR) to control very large power loads. Also note that the lamp load is shown placed on

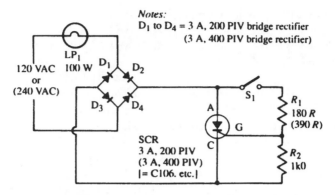

Figure 1.30 *Full-wave ON-OFF SCR circuit with AC power load*

the DC side of the bridge rectifier, and this circuit is thus shown as for use with DC loads; it can be modified for use with AC loads by simply placing the load on the AC side of the bridge, as shown in *Figure 1.30*.

Triac basics

The SCR is a unidirectional device, which can conduct current only from anode to cathode; a *triac*, on the other hand, is a bidirection device that can conduct current in either direction between its two main terminals (MT1 and MT2), and it can thus be used to directly control AC power. *Figure 1.31* shows the standard symbol of the triac, and *Figure 1.32* shows it used as a simple AC power switch that can be used to replace the *Figure 1.30* SCR-based design.

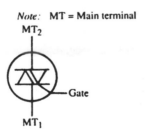

Figure 1.31 *Triac symbol*

The triac can, for most practical purposes, be regarded as a pair of SCRs wired in inverse parallel (and thus able to conduct MT2 currents in either direction) and sharing a common gate terminal. The device action is such that it can be triggered by either positive or negative gate currents, irrespective of the polarity of the MT2 current, and it thus has four possible triggering modes or 'quadrants', signified as follows:

I+ Mode = MT2 current +ve, gate current +ve.
I– Mode =   "   "   +ve, "   "   –ve
III+ Mode =   "   "   –ve, "   "   +ve
III– Mode =   "   "   –ve, "   "   –ve

Most triac data sheets carry information relating to the device sensitivity in each of the four quadrants; the trigger current sensitivity is always greatest when the MT2 and gate currents are both of the same polarity (either both positive or both negative), and is usually about half as great when they are of opposite polarity

The operation of the *Figure 1.32* circuit is thus quite simple. When S1 is open, the triac acts as an open switch and the lamp passes zero current, but when S1 is closed the triac is gated on via R1 and self-latches shortly after the start of each half-cycle, thus switching full power to the lamp load.

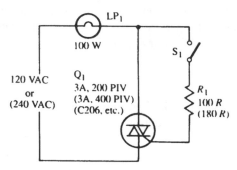

Figure 1.32   *Simple triac AC power switch circuit*

Note that SCRs and triacs can be used to apply variable power to DC or AC loads by using either the burst-fire or the phase-triggered techniques described in *Figures 1.4* and *1.5*.

SCR/triac 'rate-effect'

An SCR or triac is turned on by feeding a trigger signal to its gate. Unfortunately, internal capacitances inevitably exist between the anode and gate of an SCR or the MT1 terminal and gate of a triac, and if a sharply rising voltage is fed to an SCR/triac anode/MT1 terminal it can cause enough gate-voltage breakthrough to trigger the SCR/triac on. This unwanted 'rate-effect' turn-on can be caused by supply line transients, and sometimes occurs at the moment that supplies are switch-connected to the SCR/triac; the problem is particularly severe when driving inductive loads such as electric motors, in which load currents and voltages are out of phase.

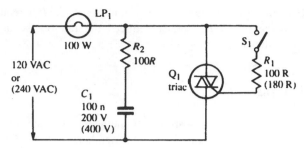

Figure 1.33 *Simple triac power switch circuit with R2–C1 'snubber' network to give rate-effect suppression*

Rate effect problems can usually be overcome by wiring an R C 'snubber' network between the anode and cathode of an SCR or MT1 and MT2 of a triac, to limit the voltage rate-of-rise to a safe value, as shown (for example) in the triac power switch circuit of *Figure 1.33*, where R2–C1 form the snubber network.

RFI suppression

Each time a resistance-driving AC-powered SCR or triac is gated on its load current switches sharply (in a few microseconds) from zero to a value determined by its load resistance and supply voltage values; this switching action inevitably generates a pulse of RFI, which is least when the device is triggered close to the 0° and 180° 'zero crossing' points of the supply line waveform (at which the switch-on currents are at their minimum), and is greatest when the device is triggered 90° after the start of each half cycle (where the switch-on currents are at their greatest).

The RFI signal magnitude is also proportional to the cable length linking the SCR/triac to its power load. Note that the RFI pulses repeat at double the supply line rate if the SCR/triac is triggered in each supply line half-cycle, and that RFI generation can thus be particularly annoying in simple lamp dimmer circuits. Fortunately, such problems can usually be eliminated by fitting the dimmer with a

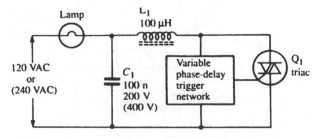

Figure 1.34 *Basic lamp dimmer circuit with RFI suppression via C1–L1*

simple L–C RFI-suppression network, as shown in *Figure 1.34*; here, the L–C filter is fitted close to the triac, and greatly reduces the rate-of-rise of supply line transition currents, thus eliminating RFI.

Diacs and quadracs

The SCR and triac are the two best-known members of a family of solid-state thyristor 'trigger' devices. Two other important members of this family are the *diac* and the *quadrac*.

Figure 1.35   *Diac symbol*

Figure 1.35 shows the standard circuit symbol of the diac, a 2-terminal bilateral trigger device that can be used with voltages of either polarity. Its basic action is such that, when connected across a voltage source via a current-limiting load resistor, it acts like a high impedance until the applied voltage rises to about 35V, at which point the diac triggers and acts like a 30V Zener diode, so 30V is developed across the diac and the remaining 5V is developed across the load resistor. The diac remains in this state until its forward current falls below a minimum holding value (this occurs when the supply voltage is reduced below the 30V 'zener' value), at which point the diac turns off again.

The diac is most often used as a trigger device in phase-triggered triac variable power control applications, as in the basic lamp dimmer circuit of *Figure 1.36*. Here,

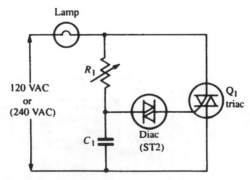

Figure 1.36   *Basic diac-type variable phase-delay lamp dimmer circuit*

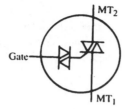

Figure 1.37 *Quadrac symbol*

in each power line half-cycle, the R1–C1 network applies a variable phase-delayed version of the half-cycle to the triac gate via the diac, and when the C1 voltage rises to 35V the diac fires and delivers a 5V trigger pulse (from C1) into the triac gate, thus turning the triac on and simultaneously applying power to the lamp load and removing the drive from the R–C network. The mean power to the load (integrated over a full half-cycle period) is thus fully variable from near zero to maximum via R1.

Some triacs are manufactured with a built-in diac in series with the triac gate; such devices are known as quadracs, and use the circuit symbol shown in *Figure 1.37*.

The unijunction transistor

The unijunction transistor (UJT) is a 3-terminal trigger device that is often used to trigger SCRs or triacs; *Figure 1.38(a)* shows its circuit symbol. It is normally used in the basic way shown in *Figure 1.38(b)*, with its B2 terminal taken to the positive supply rail and its B1 terminal grounded, and with an input signal voltage applied to its emitter (E) terminal; the action of this circuit is as follows.

Normally, when the input (E) voltage is very low, the UJT has a near-infinite E-to-B1 input impedance. If the input voltage is slowly increased a point is reached, at a 'peak-point' voltage (V_P) of about 60 percent of the supply value, at which the input impedance starts to fall and the input starts to draw a trigger current; if this current is allowed to exceed a minimum 'peak-point emitter current' (I_P) value of a few micro-amps, the UJT enters a regenerative switching phase in which the input

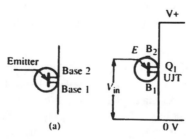

Figure 1.38 (a) *circuit symbol and* (b) *basic usage circuit of the UJT*

impedance drops to a mere 20 ohms or so. Once triggered, the UJT input impedance remains low until the input current is reduced below a 'valley-point' value (I_V) of a few milliamps, at which point the input impedance starts to switch high again.

UJT oscillator

The UJT is usually used in a relaxation oscillator circuit shown in basic form in *Figure 1.39*. Here, the UJT input is taken from the C1–R1 timer network, and the output is taken from R2, which is wired between B1 and ground. This circuit operates as follows.

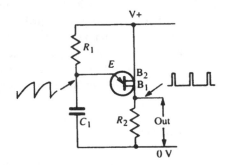

Figure 1.39   *Basic UJT relaxation oscillator circuit*

When power is first applied C1 is discharged, so E is at zero volts and has a near-infinite input impedance. As soon as power is applied, C1 starts to charge via R1 and generates an exponentially rising voltage on E; eventually, after a period determined by C1–R1, the E voltage reaches the UJT's peak-point value, and the input impedance starts to switch low. Under this condition C1 acts as a low-impedance power source, and the E-to-B1 impedance thus switches to a very low value, making C1 rapidly discharge into R2 and thus generating a brief but powerful output pulse until, eventually, the E input current falls below the UJT's valley-point value, at which point the emitter reverts to the high-impedance state, and C1 starts to recharge via R1. This process repeats *ad infinitum*, with an exponential sawtooth waveform being developed at the C1–R1 junction and a pulse waveform being developed across R2.

Note that for this circuit to work correctly the R1 value must be large enough to limit currents to less than the UJT's I_V value of a few mA (i.e. greater than a few kilohms), but small enough to allow currents to exceed the UJT's minimum I_p current of a few µA (i.e. less than 500k or so). Typically, R1 can have any value in the range 4k7 to 500k, enabling pulse delay times to be varied over a wide range via R1 and making the circuit suitable for use in a variety of phase-delayed power control applications.

Isolated-input switching

A major attraction of the UJT oscillator is that it can generate high-current output pulses (up to several hundred mA) while consuming a fairly low mean current (a couple of mA). One important application of the UJT oscillator is shown in the isolated-input AC power switch circuit of *Figure 1.40*, where the oscillator is DC powered via T2–D1–C2 and SW1 and operates at several kHz, and thus delivers roughly fifty trigger pulses to the triac gate (via isolation pulse transformer T1) during each AC power line half-cycle; consequently, the triac is triggered by the first pulse occurring in each power half-cycle, and this appears within a few degrees of the start of the half-cycle.

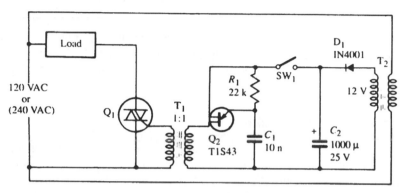

Figure 1.40   *UJT-triggered isolated-input AC power switch*

The triac is thus turned on almost permanently when SW1 is closed, and virtually full power is applied to the AC load. The trigger circuit is, however, fully isolated from the AC power source via T1 and T2, and can be turned on and off by switching a few milliamps of DC current via SW1; in practice, SW1 can easily be replaced by any type of electronic sensing circuitry, which can be fully isolated from the AC power line.

2

Switch and relay circuits

Chapter 1 took a brief look at electromechanical switches and relays, explaining their basic operating principles and showing a few simple application circuits. The present chapter expands on this theme, and shows a large selection of practical and useful application circuits.

Lamp switching circuits

The simplest power control circuit is that used to turn a filament lamp on and off. In AC line-powered applications this takes the *Figure 2.1* 'series' form, with SW1 connected to the LIVE, PHASE or 'HOT' power line and the lamp wired to the NEUTRAL or SAFE line, to minimise the consumer's chances of getting a shock when changing lamps; this simple circuit allows the lamp to be switched on and off from one point only.

Figure 2.2 shows how to switch a lamp from either of two points, by using a two-way switch at each point, with two wires (known as strapping wires) connected to each switch so that one or other wire carries the current when the lamp is turned on.

Figure 2.3 shows the above circuit modified to give lamp switching from any of three points. Here, a ganged pair of 2-way switches (SW3) are inserted in series with the two strapping wires, so that the SW1–SW2 lamp current flows directly along

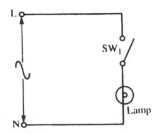

Figure 2.1 *Single-switch ON/OFF AC lamp control circuit*

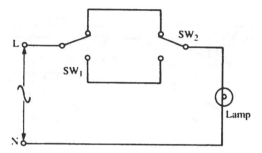

Figure 2.2 *Two-switch ON/OFF AC lamp control circuit*

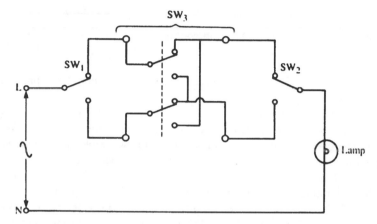

Figure 2.3 *3-switch ON/OFF AC lamp control circuit*

one strapping wire path when SW3 is in one position, but crosses from one strapping wire path to the other when SW3 is in the alternative position. Note that SW3 has opposing pairs of output terminals shorted together; in the electric wiring industry such switches are available with these terminals shorted internally, and with only four terminals externally available (as indicated by the small white circles in the diagram); these switches are known in the trade as 'intermediate' switches.

In practice, the basic *Figure 2.3* circuit can be switched from any desired number of positions by simply inserting an 'intermediate' switch into the strapping wires at each desired new switching position; *Figure 2.4*, for example, shows the circuit modified for four-position switching.

Multi-input switching circuits

Note that although the *Figure 2.4* circuit is electrically simple, it may be quite difficult to install physically, since each intermediate switch needs at least a 3-core (LIVE, NEUTRAL, and GROUND) heavy duty cable feeding into it and a similar

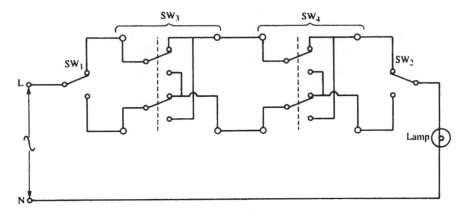

Figure 2.4   *4-switch ON/OFF AC lamp control circuit*

cable feeding out, and these must be buried in channels cut into plaster and masonry if concealed wiring is wanted.

These installation problems can be minimised in any of three basic ways. One technique often used in long corridors in hotels and guest houses, etc., where lights need to be switched from many different points, is to use simple push-operated timer switches which stay closed for only a minute or so once operated, connected in parallel as in *Figure 2.5*, thus enabling the switch wiring to be greatly simplified.

Another technique is to use light-duty multi-way switching circuitry (with wiring that can easily be hidden in plaster, etc.) to activate a low-voltage relay (powered from an AC 'mains'-derived 12 volt DC supply), which then switches the AC-powered lamp via one set of relay contacts, as shown (for example) in the 2-way switching circuit of *Figure 2.6*.

Finally, the best technique of all is to switch the lamp via a relay that is electronically activated via multi-input push-button control circuitry, thus enabling very simply light-duty switch wiring to be used. Practical circuits of this type are described later in this chapter (see *Figure 2.19*).

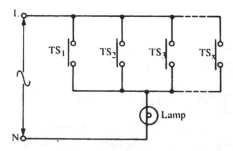

Figure 2.5   *Multi-input AC lamp switching circuit using simple timer switches*

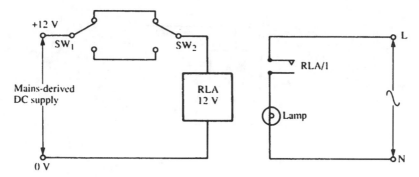

Figure 2.6 *Multi-input relay-activated AC lamp switching circuit*

Relay configurations

The simplest way of using a relay is in the basic non-latching mode, with the activating switch in series with its coil as already shown in *Figure 1.10*, so that the relay closes only when the switch is closed.

In practice, a relay may be activated via several switches, either wired in series to give AND-logic operation, or in parallel to give OR-logic operation, as shown in *Figures 2.7* and *2.8*; in the AND-logic circuit the relay turns on only when *all* series-connected switches (SW1 AND SW2 AND SW3, etc) are closed at the same time; in the OR-logic circuit the relay turns on when *any* of the parallel-connected switches (SW1 OR SW2 OR SW3, etc.) are closed.

Coil damping

Relay coils are inductive, and may generate back-e.m.f.s of hundreds of volts if their coil currents are suddenly broken. These back-e.m.f.s can easily damage switch contacts or solid-state devices connected to the coil, and it is thus often necessary

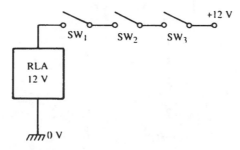

Figure 2.7 *AND logic relay switching*

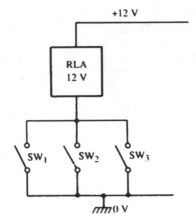

Figure 2.8   *OR logic relay switching*

to 'damp' these coil back-e.m.f.s via protective diodes; *Figures 2.9* and *2.10* show examples of such circuits.

In *Figure 2.9* the coil damping is provided via D1, which prevents switch-off back-e.m.f.s from driving the RLA–SW1 junction more than 600mV above the positive supply rail value. This form of protection is adequate for many practical applications.

In *Figure 2.10* the damping is provided via two diodes that stop the RLA–SW1 junction from swinging more than 600mV above the positive supply rail or below the zero-volt rail. This form of protection is adequate for even the most critical needs, and is recommended for all applications in which SW1 is replaced by a transistor or other 'solid-state' switch.

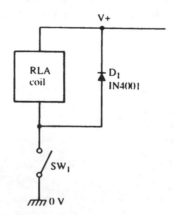

Figure 2.9   *Single-diode relay coil damper*

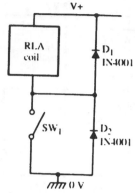

Figure 2.10 *Two-diode relay coil damper*

High-Z relay switches

Normal and reed-type relays are low-impedance devices, with typical coil impedances of only a few tens or hundreds of ohms. Their effective input impedances can easily be increased (so that they act as high-impedance or 'high-Z' relay switches), however, by simply interposing a transistor or IC driver stage between the relay coil and the input drive signal. *Figures 2.11* to *2.14* show four variations of this theme.

In *Figure 2.11*, Q1 is wired as a common-emitter amplifier and increases the effective relay coil-current sensitivity by a factor of about ×100 (=the current gain of Q1) and also increases the voltage sensitivity to a few volts. R1 limits the Q1 input current to a safe value and also dictates the effective input impedance of the circuit (=R1 plus roughly 1k0). D1 and D2 damp the coil back-e.m.f.s. The RLA/1 contacts can control external circuitry.

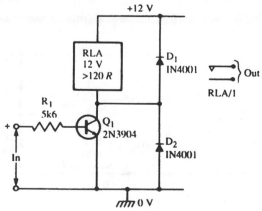

Figure 2.11 *Non-latching transistor-driven relay switch*

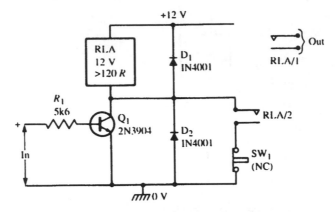

Figure 2.12   *Self-latching transistor-driven relay switch*

The *Figure 2.11* circuit gives a non-latching action in which RLA is OFF when the input is below 600mV and is ON when the input exceeds a few volts. The circuit can be made to give a self-latching action by modifying it as in *Figure 2.12*, where n.o. contacts RLA/2 (plus n.c. switch SW1) are in parallel with Q1 and thus short Q1 and self-latch RLA once it has been initially activated. Once it has self-latched, the relay can only be turned off again by either opening SW1 or breaking the supply-line connections.

The above two circuits have input impedances of a few kilohms; if desired, impedance of 10M or more can be obtained by driving Q1 via a CMOS buffer made from one or more stages of a 4001B quad 2-input NOR gate wired in the inverter mode (by shorting the gate input pins together), as shown in *Figures 2.13* and *2.14*.

In *Figure 2.13* only a single CMOS inverter stage is used; consequently, to ensure that the relay turns off when the input voltage is low, the Q1 driver must be a pnp device. In *Figure 2.14*, two CMOS inverter stages are wired in series, to give zero overall signal inversion, and Q1 can thus by an npn device. In these two circuits, the

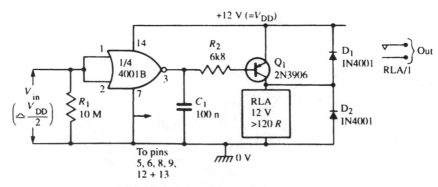

Figure 2.13   *Simple high-impedance relay switch*

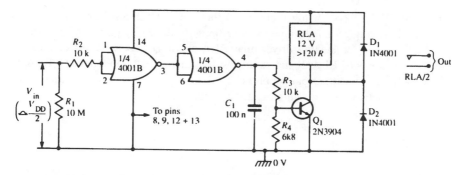

Figure 2.14   *Modified high-impedance relay switch*

input impedance equals the R1 value, and the relay turns ON (or OFF) when the input signal goes above (or falls below) the half-supply-voltage (approx.) 'transition' voltage value of the CMOS input gate, at which value the gate operates in the linear mode; C1 inhibits any high frequency or transient instability that may occur in the gate when in this linear mode.

A simple burglar alarm

Figure 2.15 shows a simple but useful relay-output burglar alarm that consumes a mean current of only 1µA when in the 'standby' mode (with SW1 and all series-

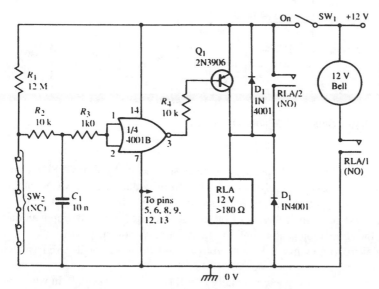

Figure 2.15   *Simple burglar alarm, activated by series-connected n.c. switches*

connected SW2 sensor switches closed). When any of the SW2 sensor switches open, a 'high' voltage is fed to the input of the 4001B CMOS gate, which is wired as an inverting buffer and thus drives relay RLA on via Q1. As RLA turns on it self-latches via contacts RLA/2 and activates the alarm bell via contacts RLA/1. R2–R3–C1 act as a transient-suppressing filter network that protects the alarm against false activation via lightning strikes, etc.

Bistable circuits

A relay can be made to give a bistable action in which it turns on and self-latches when a SET push-button switch is operated, and can then subsequently be turned off again only by operating a RESET push-button switch, by using the connections shown in *Figure 2.16*. Here, two NOR gates (taken from a 4001B CMOS IC) are wired as a simple manually activated bistable multivibrator that has one of its outputs taken to the relay coil via the Q1 common-emitter buffer stage.

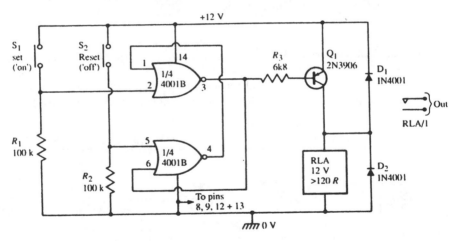

Figure 2.16 *Simple bistable relay switch*

The *Figure 2.16* circuit actually changes state as its SET or RESET input signal rises through the half-supply-voltage 'transition' value of the CMOS gate, and this fact enables the circuit to be modified to give a self-latching relay action as in *Figure 2.17*; here, the relay turns on and self-latches when the input voltage rises above the 'transition' value, and the relay can subsequently be turned off only by removing (or reducing) the input voltage and pressing the RESET switch. This circuit has an input impedance of 10M.

Figure 2.18 shows an exceptionally useful type of relay switch, in which the relay can be turned on and off via a single push-button switch, or by any number of such

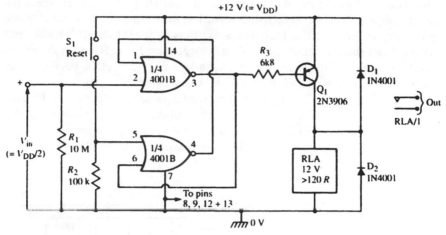

Figure 2.17 *Self-latching high-impedance relay switch*

switches wired in parallel, thus enabling the relay to be remote controlled from any number of input points. The circuit action is such that the relay changes state each time an input switch is operated (pressed and released): thus, if the relay is initially ON, it will turn OFF when any switch is next operated, and ON again when any switch is operated after that, and so on. The circuit thus gives a 'binary' relay switching action.

The *Figure 2.18* circuit is designed around a 4013B CMOS dual D-type flip-flop, with one flip-flop disabled by taking its inputs to ground and the other configured

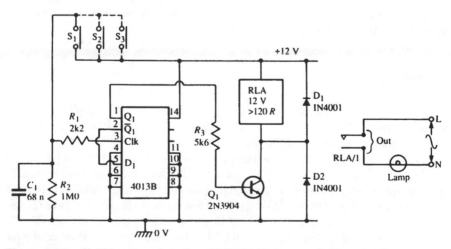

Figure 2.18 *Multi-input push-button ON/OFF AC lamp switching circuit*

as a divide-by-2 circuit (by shorting its NOT-Q and D1 pins together). The input clock pulses to this divider need rise times of less than 15μs, and these are obtained via the push-button switches; each time a switch is closed C1 charges rapidly, thus giving the fast-rise-time clock pulse; C1 discharges slowly via R2 when the switch reopens, eliminating false triggering via switch-bounce effects, etc. Q1 and the relay thus reliably change state each time a push-button switch is operated.

Circuits of the *Figure 2.18* type are of particular value in enabling hall, landing, or corridor lights to be controlled from several different switching points. In this case the push-button control switches can be connected to the main unit via very thin twin cable that can easily be hidden from sight. In this application the unit

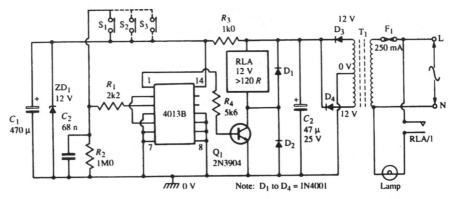

Figure 2.19 *Mains powered version of the push-button lamp switching circuit*

should (naturally) be powered from the AC mains (power) line, and *Figure 2.19* shows how it can be modified for this type of operation. Mains transformer T1 must provide a 12V–0–12V output at 100mA or greater.

Timer circuits

Relays can be used in many time-delayed-switching or timer applications, and *Figures 2.20* to *2.26* show some practical examples of circuits giving delays varying from a fraction of a second up to tens of hours.

Figures 2.20 to *2.22* show how inexpensive 4001B CMOS ICs can be used to implement medium-accuracy time delays of up to several minutes. The *Figure 2.20* design gives a delayed turn-on relay switching action. Here, the CMOS gate is wired as a simple inverter, with its output driving the relay coil via pnp transistor Q1 and with its input taken from the junction of time-controlled potential divider R2–C1. When power is first applied to the circuit C1 is fully discharged, so the inverter input is grounded, its output is at full positive-rail potential, and Q1 and the relay are off.

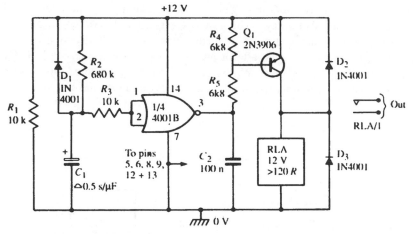

Figure 2.20 *Delayed-turn-on relay switch*

As soon as power is applied C1 starts to charge up via R2 and applies a rising exponential voltage to the inverter input until, after a delay determined by the C1–R2 values, it reaches the inverter's threshold value, forcing its output to swing downwards and drive Q1 and the relay on. The relay then remains on until circuit power is removed, at which point C1 discharges rapidly via D1 and R1; the operating sequence is then complete.

Figure 2.21 shows how the above circuit action can be reversed, so that the relay turns on as soon as power is applied but then turns off again after a pre-set delay, by simply modifying the relay-driving stage for npn transistor operation.

The above two circuits each give a delay of about 0.5secs/μF of C1 value, enabling delays up to several minutes to be obtained; delays can be made variable by using a variable resistor in the R2 position.

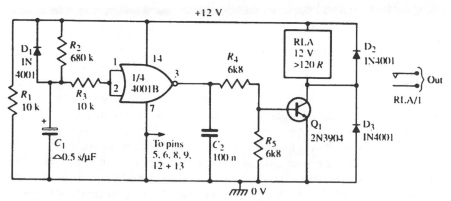

Figure 2.21 *Auto-turn-off relay switch*

Figure 2.22 shows a pair of CMOS gates used to make a push-button-activated medium-accuracy one-shot relay switch that gives time delays up to several minutes. The action is such that the relay turns on as soon as START switch S1 is briefly closed, but turns off again after a pre-set delay of about 0.5sec/μF of C1 value. The two CMOS gates are simply wired as a manually triggered monostable multivibrator that has its output fed to the relay via R4 and Q1, and has its timing controlled by C1–R3.

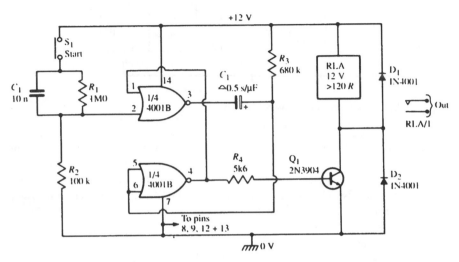

Figure 2.22   *One-shot relay time switch*

Note in the circuits of *Figures 2.20* to *2.22* that C1 *must* be a low-leakage component, with a leakage resistance far greater than 1.5 megohms. If any of these circuits fail to work, C1 is probably defective.

555 timer circuits

The *Figure 2.20* to *2.22* relay-driving power-switching circuits are all based on simple CMOS elements, and are intended for use in medium-accuracy timer applications only. Far greater timing accuracy can be obtained by using a popular type-555 timer IC as the basic circuit timing element, and *Figures 2.23* to *2.26* show four practical high-accuracy relay-driving timer circuits that are designed around this IC.

Figure 2.23 shows a simple 6-second to 60-second timer in which the 555 is wired in the monostable or one-shot mode. A timing cycle is started by briefly closing S1; relay RLA immediately turns on and C1 starts to charge towards the positive rail via R2 and RV1 until eventually, after a delay determined by the RV1

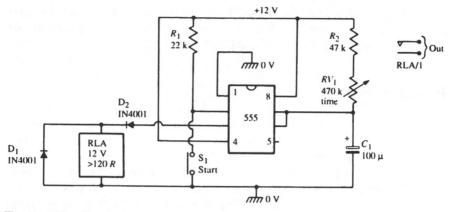

Figure 2.23 *Simple 6 to 60 seconds timer circuit*

setting, C1 rises to two-thirds of the supply rail voltage, at which point the IC changes state and the relay turns off. The timing cycle is then complete.

Note that the above circuit draws supply current even when the relay is off; *Figure 2.24* shows a 2-range timer circuit that does not suffer from this defect, and which covers the timing range 6 seconds to 10 minutes. Here, when S1 is briefly closed a START pulse is fed to pin-2 of the IC via R1 and C1, and the relay turns on; contacts RLA/2 then change over and maintain the power connections when S1 is released. The circuit then runs through a timing cycle like that described above, but with its timing period determined by either C2 or C3, until eventually the relay

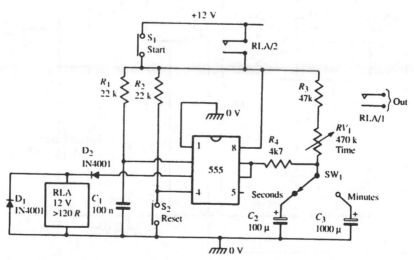

Figure 2.24 *2-range 6 to 60 seconds and 1 to 10 minutes timer*

turns off and contacts RLA/2 reopen and break the circuit's supply connections. The timing cycle is then complete. Note that the circuit can be turned off part way through its timing cycle by operating RESET switch S2.

Precision timers

Conventional electrolytic capacitors have very wide tolerance values (typically –50 percent to +100 percent), and have fairly large and unpredictable leakage currents. Consequently, simple circuits of the types shown in *Figures 2.23* and *2.24* can not be relied on to give precise timing periods or to give periods that exceed 15 minutes or so. *Figures 2.25* and *2.26* show two high-accuracy long-period timers that use non-electrolytic capacitors in their timing networks, and thus provide inherently excellent reliability.

In these two circuits IC1 is wired as a stable, variable frequency, free-running astable multivibrator. In the *Figure 2.25* design the astable frequency is divided

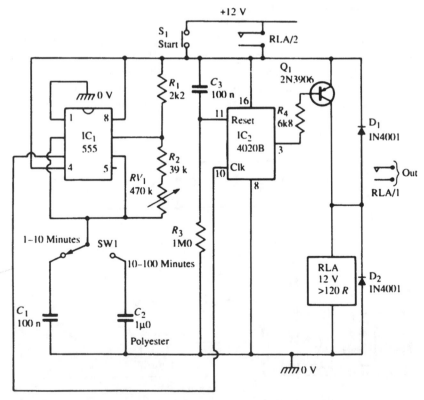

Figure 2.25   *2-range 1 to 10 minutes and 10 to 100 minutes timer*

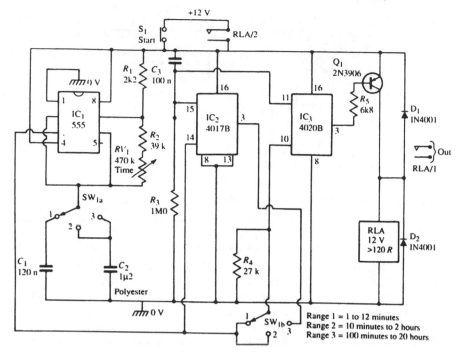

Figure 2.26 *Wide-range timer spans 1 minute to 20 hours in three ranges*

down by IC2, a 14-stage binary divider, so that the relay turns on as soon as S1 is closed, but turns off again on the arrival of the 8192nd astable pulse, thereby giving total timing periods in the range 1 to 100 minutes.

The *Figure 2.26* circuit is similar, except that an additional decade-divider stage is used in position-3 of SW_1, thus giving a maximum division ratio of 81 920 and making maximum timing periods of up to 20 hours available from the unit. This circuit is of particular value in giving time-controlled turn-off of battery chargers, etc.

Relay pulser circuits

A relay pulser is a circuit in which the relay is repeatedly driven on and off for pre-set periods. *Figure 2.27* shows a simple version of such a circuit, wired as a lamp-flasher. Circuit operation relies on the fact that there is always a substantial difference between a relay coil's *pull-in* voltage (the value that just makes the contacts close) and its *drop-out* voltage (at which the contacts reopen). Typically, a 12V relay may pull-in at about 10V and drop-out at about 5V; thus, the *Figure 2.27* circuit operates as follows.

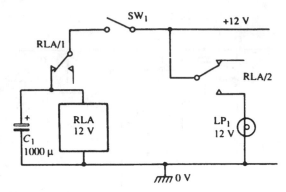

Figure 2.27   *Relay pulser or lamp flasher circuit*

When SW1 is first closed C1 charges rapidly via closed contacts RLA/1 until the relay coil voltage reaches its pull-in value, at which point contacts RLA/1 open. C1 then starts to discharge into the relay coil and thus holds the RLA/1 contacts open until the C1 voltage falls to the relay's drop-out value, at which point contacts RLA/1 close again, and the whole timing process starts to repeat. Thus, the relay contacts repeatedly open and close at a rate determined by the C1 and coil-resistance values and by the relative values of coil pull-in and drop-out voltages. The lamp (or other external circuitry) is activated via contacts RLA/2.

This simple relay pulser gives fairly poor timing accuracy, and calls for the use of a large value electrolytic timing capacitor. *Figure 2.28* shows a precision relay pulser that operates at a 1Hz rate and uses a 10μF timing capacitor. This design is based on a modified version of the standard '555 timer IC' astable multivibrator circuit.

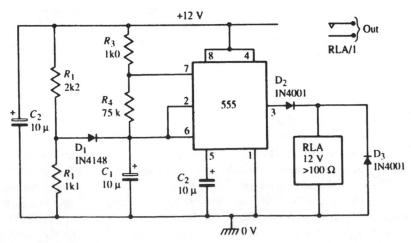

Figure 2.28   *'Precision' 1 Hz relay pulser*

In the standard form of this astable, C1 first charges to two-thirds of the supply-voltage value via R3–R4, at which point the 555 changes state and C1 then discharges to one-third of the supply-voltage value via R4, at which point the 555 reverts to its original state and C1 starts to recharge again via R3–R4, and so on *ad infinitum*. Thus, once the initial 'start-up' cycle is over, the C1 voltage repeatedly swings between one-third and two-thirds of the supply voltage value, and the IC generates a nearly-symmetrical squarewave output that switches the relay on and off at a 1 Hz rate. Note, however, that in the first 'start-up' half-cycle C1 has to charge from zero to two-thirds of the supply voltage, and this half-cycle is thus much longer than the succeeding ones. In *Figure 2.28* this fault is overcome via R1–R2 and D1, which rapidly charge C1 to one-third of the supply voltage during the start-up half-cycle, thus giving a near-equal or 'precision' timing action on all operating half-cycles.

Photo-sensor basics

Relay switching circuits can easily be coupled to a photo-sensor and made to activate when light intensity goes above or below a pre-set value, or when a light beam is broken by a person or object, or when a light source is reflected onto the face of a light sensor by particles of smoke or fog, etc. The best-known and easiest-to-use type of light-sensitive device (photo-sensor) is the LDR or 'light sensitive resistor', which uses the symbol shown in *Figure 2.29(a)*.

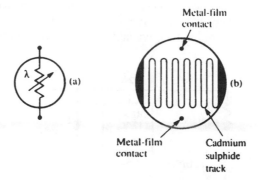

Figure 2.29 *LDR symbol* (a) *and basic structure* (b)

LDR operation relies on the fact that the resistance of a cadmium sulphide (CdS) film varies with the intensity of light falling on its face; the resistance is very high under dark conditions, and low under bright conditions. *Figure 2.29(b)* shows the LDR's basic construction, which consists of a pair of metal film contacts separated by a snake-like track of cadmium sulphide film designed to give the maximum possible contact area with the two metal films. The structure is housed in a clear plastic or resin case that gives free access to external light.

Practical LDRs are sensitive, inexpensive and readily available devices with good voltage and power handling capabilities, similar to those of a normal resistor. They are available in a several sizes and package styles, the most popular size having a face diameter of about 10mm. Typically, such a device has a resistance of several megohms under dark conditions, falling to about 900R at a light intensity of 100 lux (typical of a well-lit room) or about 30R at 8000 lux (typical of bright sunlight).

Figures 2.30 to *2.36* show a selection of practical light-sensitive relay-output switching circuits that use LDRs as their opto-sensors; each of these circuits will work with virtually any LDR with a face diameter in the range 3mm to 12mm.

Light-activated relays

Figures 2.30 to *2.32* show a selection of relay output light-activated 'switch' circuits based on the LDR. *Figure 2.30* shows a simple non-latching circuit that activates when light enters a normally dark area such as the inside of a safe or cabinet, etc.

Here, R1–LDR and R2 form a light-sensitive potential divider that determines the base bias of Q1. Under dark conditions the LDR resistance is high, so zero base bias is fed to Q1, and Q1 and the relay are off. When a significant amount of light falls on the LDR face its resistance falls to a fairly low value, applying bias to the base of Q1, which thus turns on and activates the RLA/1 relay contacts, which can be used to control external circuitry.

The simple *Figure 2.30* circuit has a fairly low sensitivity, and its 'trigger' points are fairly susceptible to changes in supply voltage and ambient temperature. *Figure 2.31* shows a CMOS-aided light-operated relay circuit that does not suffer from this susceptibility. Here, one of the four available NOR gates of a 4001B CMOS IC is

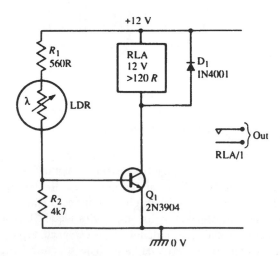

Figure 2.30   *Simple non-latching light-activated relay switch*

used as a linear inverter between the light-sensitive LDR–RV1 network and the input of relay-driving transistor Q1.

When the 4001B CMOS NOR gate is wired as an inverter it gives linear action only when its input is within a few dozen millivolts of the inverter's 'threshold voltage' point; at all other times the gate output is saturated (driven to either ground or positive supply voltage). This 'threshold' point is actually a fixed fraction of the supply voltage value (the value is typically 50 percent, but may vary from 30 percent to 70 percent between individual devices), and is very stable.

Thus, when it is wired in the configuration shown in *Figure 2.31*, the inverter input acts like the 'fixed potential divider' side of a Wheatstone bridge (with LDR–RV1 forming the 'variable' side of the bridge) and acts as a bridge-balance detector that goes into the linear mode when the bridge is very close to its balance point, which is not greatly influenced by variations in supply voltage or ambient temperature.

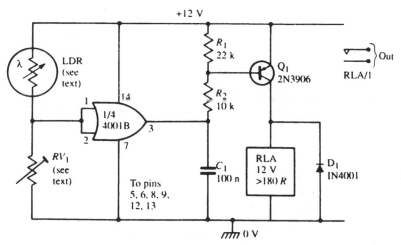

Figure 2.31 *CMOS-aided light-operated switch*

The action of this circuit is such that the gate input is low when the light level is below the desired 'trip' value, thus driving the gate output high and turning Q1 and RLA off; but when the light level is above the trip value the gate input is high and its output is driven low, thus driving Q1 and RLA on. When the light level is very close to the desired trip level the gate is driven into the linear mode, and small changes in light level can make the relay switch on or off; the trip level can be pre-set via RV1.

The action of the above circuit can be reversed, so that the relay goes on when light intensity falls below a pre-set level, by transposing the RV1 and LDR positions as in *Figure 2.32*, which also shows how a simple transient-suppressing network can be wired between the output of the light-sensitive divider and the input of the CMOS

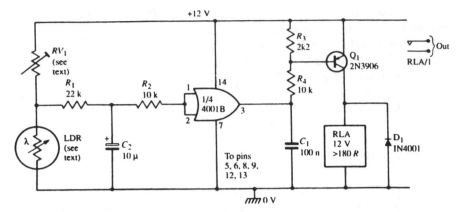

Figure 2.32   *CMOS-aided dark-operated switch with transient suppression*

gate, so that the circuit responds to mean light levels but is unaffected by sudden light transients (such as are caused by lightning flashes, etc.). This circuit can be used to automatically turn porch lights or car parking lights on at dusk and off at dawn.

Note in the above two circuits that the LDR can be any type that gives a resistance in the range 2k0 to 2M0 at the desired 'trip' level, and that (when adjusted) the RV1 value should balance that of the LDR. C1 is used to ensure stability of the CMOS inverter when it is operating in the linear mode.

Precision light-activated circuits

The CMOS-aided circuits of *Figures 2.31* and *2.32* give a semi-precision light-sensitive switching action that is adequate for most practical purposes. If even greater precision is needed it can be obtained by using the op-amp circuits of *Figures 2.33* to *2.35*.

In *Figure 2.33*, LDR–RV1 and R1–R2 form a light-sensitive Wheatstone bridge that has its output fed to a sensitive 741 op-amp 'balance' detector; R1–R2 feed a fixed half-supply voltage to the detector's non-inverting input, and LDR–RV1 feed a light-dependent voltage to its inverting input. If these two voltages differ by more than a few mV the op-amp output is driven to saturation (to near-zero or near-positive-rail values), thus driving Q1 and RLA either on or off. If the two voltages are within a few mV of each other, the relay state depends on the direction of bridge imbalance. The actual balance point can be pre-set via RV1, and is independent of variations in supply voltage or temperature. Because of the very high gain of the op-amp, the *Figure 2.33* circuit has a far greater sensitivity than the CMOS-aided circuits described earlier.

The *Figure 2.33* circuit is configured so that the relay turns on when the light goes above a pre-set level. The action can be reversed, so that it acts as a 'dark-activated'

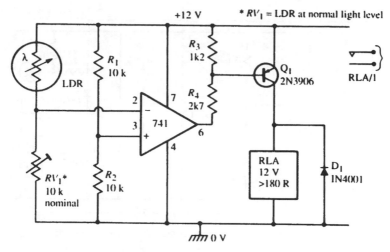

Figure 2.33 *Precision light-activated relay switch*

switch, either by transposing the inverting and non-inverting input connections of the op-amp, or by transposing the LDR and RV1, as shown in *Figure 2.34*, which also shows how a small amount of hysteresis can be added via feedback resistor R5, so that the relay turns on when the light level falls to a particular value but does not turn off again until it rises a substantial amount above this value. The hysteresis value is inversely proportional to the R5 value, and is zero when R5 is open circuit.

A precision combined light/dark switch, that activates a single relay if the light level goes above one pre-set value or below another, can be made by combining

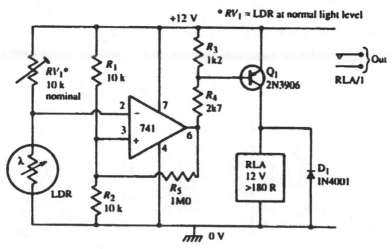

Figure 2.34 *Precision dark-activated switch with hysteresis*

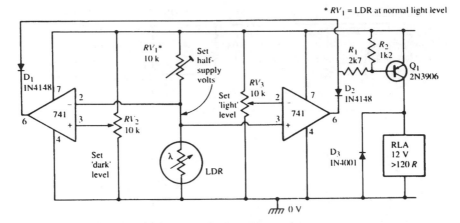

Figure 2.35   *Combined light-/dark-activated switch with single relay output*

op-amp 'light' and 'dark' switches in the manner shown in *Figure 2.35*. To set up this circuit, first adjust RV1 so that roughly half-supply volts appear on the RV1–LDR junction when LDR is illuminated at the 'normal' or mean level. RV2 can then be set so that RLA turns on when the light falls below the desired 'dark' level, and RV3 can be set so that RLA turns on when the light rises above the desired 'light' or brightness level.

'Light-beam' relay switches

One popular application of the LDR is as a light-beam alarm or switch, which activates when the passage of a beam of light is interrupted by an object or person. *Figure 2.36* shows a simple circuit of this type; the two lenses focus the lamp-

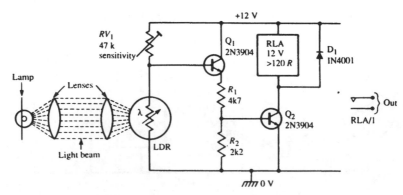

Figure 2.36   *Simple light-beam alarm with relay output*

generated light beam onto the LDR face, and the LDR circuit acts like a 'dark-operated' relay switch. Normally, with the light beam uninterrupted, the LDR face is illuminated by the beam and presents a low resistance, so little voltage appears at the RV1–LDR junction and Q2 and the relay are off. When the light beam is broken the LDR resistance increases, so a significant voltage appears on the RV1–LDR junction and activates the relay via Q2.

In practice, most modern 'beam' alarms use a modulated infra-red (IR) signal to generate the 'light beam', which is invisible, and use one or more infra-red photo diodes or transistors to detect the beam. Some simple circuits of this type are presented in the author's *Optoelectronics Circuits Manual*.

LDR-based 'smoke' alarms

Most modern low-cost 'smoke' or 'fire' alarms incorporate a low-level radioactive element that responds to the presence of ionised particles of the types found in most types of smoke and gaseous fumes. There is, however, a different type of 'smoke' alarm, which uses an LDR as its 'sensing' element and activates when smoke causes a light source to reflect onto the LDR face. *Figure 2.37* shows a sectional view of a

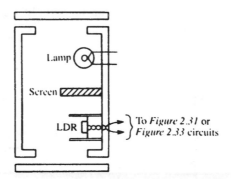

Figure 2.37 *Sectional view of reflective-type smoke detector*

reflective-type smoke detector, which consists of a lamp and an LDR mounted in an open-ended but light-excluding box with an internal screen that stops the lamplight falling directly onto the LDR face. The lamp is a source of both light and heat, and the heat causes convection currents of air to be drawn in from the bottom of the box and to be expelled through the top. The inside of the box is painted matt black, and the construction lets air pass through the box but excludes external light.

Thus, if the convected air currents are smoke free no light falls on the LDR face, and the LDR presents a high resistance, but if smoke is present its particles make the lamplight reflect onto the LDR face and so cause a great and easily detectable decrease in the LDR resistance. The detector circuitry thus needs to act like a 'light-operated' circuit, and can take the form shown in *Figure 2.31* or *2.33*. Reflective smoke detectors of this type are far less susceptible to false alarms that conventional 'ionisation' types of smoke alarm.

Heat-sensitive circuits

Each of the light-sensitive switching circuits of *Figures 2.30* to *2.36* can be converted into a temperature-sensitive switch by replacing its LDR with a negative-temperature-coefficient (n.t.c.) thermistor. These are simple devices that present a resistance value inversely proportional to temperature, i.e. resistance falls as temperature rises, and vice versa.

Thus, *Figure 2.38* shows how the precision light-activated relay switch of Figure 2.33 can be converted into a precision over-temperature switch. The thermistor used here can be any n.t.c. type that presents a resistance in the range 1k0 to 20k at the required trigger temperature, and the RV1 resistance should equal this value at the same temperature. The above circuit can be made to act as an 'ice' or under-temperature switch by simply transposing TH1 and RV1.

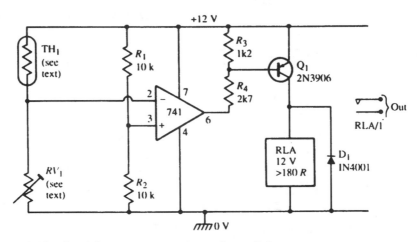

Figure 2.38   *Precision over-temperature relay switch*

Ordinary silicon diodes can be used as thermal sensors; they generate a volt drop of about 600mV at a forward current of 1mA and if this current is held constant the volt drop changes by about −2mV for each degree centigrade increase in diode temperature; all silicon diodes have inherently similar thermal characteristics; *Figure 2.39* shows how such a diode can be used as a sensor in an over-temperature switch.

Here, ZD1–R1 generate a constant 5.6V across the R2–RV1 and R3–D1 potential dividers, causing a virtual constant current to flow in each divider. A constant reference voltage is thus developed between the ZD1–R1 junction and pin-2 of the op-amp, and a temperature-dependent voltage with a coefficient of −2mV/°C is

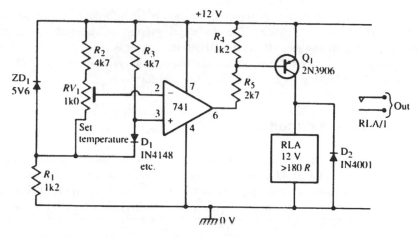

Figure 2.39 *Over-temperature switch with silicon diode sensor*

developed between the ZD1–R1 junction and pin-3 of the op-amp. Thus, a differential n.t.c. voltage appears between pins 2 and 3 of the op-amp.

To set up this circuit, simply apply the desired 'trip' temperature to D1 and set RV1 so that RLA just turns on. The circuit has an 'ON/OFF' sensitivity of about 0.5°C, and can be used as an over-value switch at temperatures ranging from sub-zero to above the boiling point of water. Note that the circuit operation can be reversed, so that it acts as an under-temperature switch, by simply transposing the pin-2 and pin-3 connections of the op-amp.

Finally, *Figure 2.40* shows two silicon diodes used as thermal sensors in a differential-temperature switch that turns on only when the D2 temperature is more

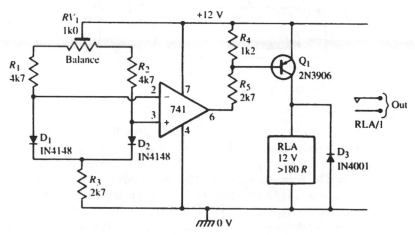

Figure 2.40 *Differential temperature switch*

than a pre-set amount greater than that of D1, and is not influenced by the absolute temperatures of the two diodes. The magnitude of this differential trip is fully variable from zero to about 10°C via RV1, so the circuit is quite versatile; it is set up by simply raising the D2 temperature the required amount above that of D1 and then carefully adjusting RV1 so that the relay just turns on.

Miscellaneous circuits

To complete this chapter, *Figures 2.41* to *2.43* show three miscellaneous types of relay-switching circuit. The *Figure 2.41* design is that of another simple relay pulser, which repeatedly switches the relay on and off at a rate variable (via RV1) between 26 and 80 cycles per minute via the CMOS astable multivibrator circuit built around the two 4001B NOR gates.

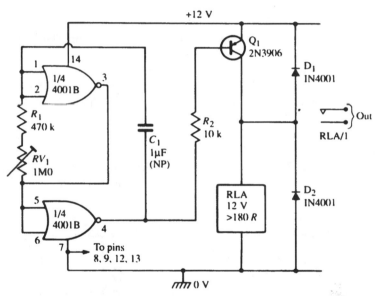

Figure 2.41   *Relay pulser circuit*

   The *Figure 2.41* circuit can be used as an emergency lamp flasher by using its relay contacts to switch power to the lamps. *Figure 2.42* shows how it can be modified so that it starts pulsing or 'flashing' automatically when the ambient light level falls below a level pre-set via RV2. Note that if this circuit is used as a lamp flasher care must be taken to ensure that the LDR face points away from the lamps, so that a photo-coupled feedback loop is not set up.
   Finally, *Figure 2.43* shows a water-activated relay switch that can be used to sound an alarm (via the relay contacts) or cut off the water supply (via a relay-activated

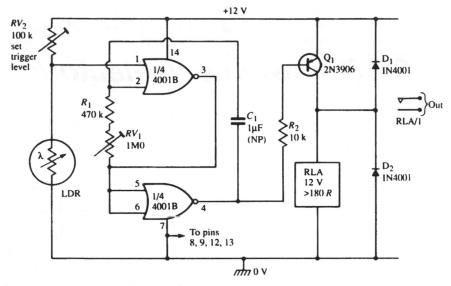

Figure 2.42   *Light-activated relay pulser*

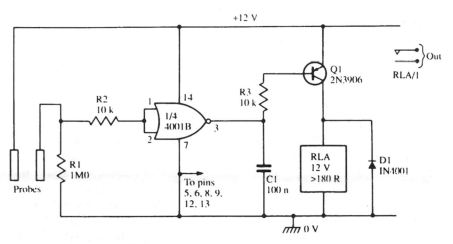

Figure 2.43   *Water-activated relay switch*

solenoid) when water reaches a pre-set level in a tank, bath, or other container. When the two metal probes are placed in water (or any other conductive liquid) the potential divider action of R1 and the liquid resistance applies a 'high' voltage to the input of the inverter-connected CMOS NOR gate, driving its output low and activating RLA via Q1. When the probes are not in a conductive liquid the gate's input is grounded via R1, and the relay is off.

3

CMOS switches/selectors

The basic operating principles of CMOS bilateral switches and multiplexer/demultiplexer ICs were briefly explained in Chapter 1, which also explained that these devices can be used to replace conventional switches and relays in many low-level power control applications. The present chapter expands on this theme and looks at a variety of practical circuits of these types.

CMOS bilateral switches

A CMOS bilateral switch or 'transmission gate' can be regarded as a near-perfect single-pole single-throw (SPST) electronic switch which, unlike an ordinary transistor or FET switch (which can conduct current in one direction only) can conduct analogue or digital signal currents in either direction (bilaterally) and can be turned on (closed) or off (opened) by applying a logic-1 or logic-0 signal to a single control terminal. Practical versions of such switches have a near-infinite OFF impedance, and a typical ON impedance in the range 90 to 300 ohms.

Standard CMOS bilateral switches can be switched at frequencies ranging from zero to several MHz, and have many uses in low-level power control circuits. They can, for example, be used to replace conventional mechanical switches and relays in signal-carrying applications, the bilateral switch being dc controlled and placed directly on the PCB where it is needed, thus eliminating the problems of signal radiation and interaction that normally occur when such signals are mechanically switched via lengthy cables.

At high frequencies, CMOS bilateral switches can be used in such diverse applications as signal gating, multiplexing, A-to-D and D-to-A conversion, digital control of frequency, impedance and signal gain, the synthesising of multi-gang pots and capacitors, and the implementation of sample-and-hold circuits. Examples of most of these applications are shown later in this chapter.

Several types of CMOS 'multiple bilateral switch' ICs are available. These range from the simple 4016B and 4066B types, which each house four independently accessible SPST bilateral switches, to the 4097B, which houses an array of bilateral switches and logic networks arranged in the form of two independently accessible single-pole 8-way bilateral switches or multiplexers/demultiplexers. Before taking a closer look at the range of ICs, however, it is wise to first examine the basic operating principles and terminology, etc., of the CMOS bilateral switch.

Basic operating principles

Figure 3.1 shows *(a)* the basic circuit and *(b)* the equivalent circuit of a simple CMOS bilateral switch. Here, an n-channel and p-channel MOSFET are wired in inverse parallel (drain-to-source and source-to-drain), but have their gate bias applied in antiphase from the control terminal via a pair of CMOS inverter stages, to give the 'bilateral' switching action. Thus, when the control signal is at the logic-0 level the gate of Q2 is driven to logic-1 and the gate of Q1 is driven to logic-0,

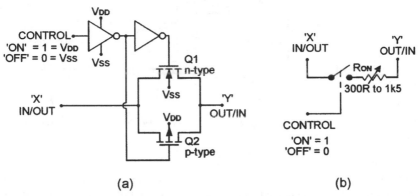

(a)   (b)

Figure 3.1 *Basic circuit* (a) *and equivalent circuit* (b) *of the simple CMOS bilateral switch*

and under this condition both MOSFETs act as open switches between the circuit's X and Y points. When, on the other hand, the control signal is set to the logic-1 level, Q2's gate is driven to logic-0 and Q1's gate is driven to logic-1, and under this condition both MOSFETs act as closed switches, and a low resistance (equal to the R_{SAT} value) exists between the X and Y points.

Note that, when the control input is at logic-1, signal currents can flow in either direction between the X and Y terminals (via Q1 in one direction and Q2 in the other), provided that the signal voltages are within the V_{SS}-to-V_{DD} power supply voltage values. Each of the X and Y terminals can thus be used as either an IN or OUT signal terminal.

In practice, Q1 and Q2 have a finite resistance (R_{SAT}) when driven on, and in this simple circuit the R_{SAT} value may vary from 300R to 1k5, depending on the magnitude of the circuit's supply voltage and on the size and polarity of the actual input signal. This simple bilateral switch can thus be represented by the equivalent circuit of *Figure 3.1(b)*.

Figure 3.2 shows an improved version of the CMOS bilateral switch, together with its equivalent circuit. This circuit is similar to the above, except for the addition of a second bilateral switch (Q3–Q4) that is wired in series with Q5, with the 'well' of Q1 tied to Q5 drain. These modifications cause Q1's well to switch to V_{SS} when the Q1–Q2 bilateral switch is off, but to be tied to the 'X' input terminal when the switch is on. This modification reduces the ON resistance of the Q1–Q2 bilateral

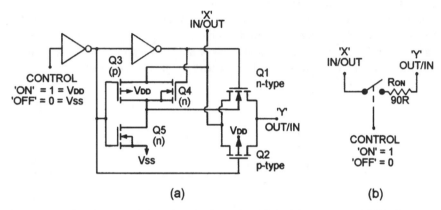

Figure 3.2   *Basic circuit (a) and equivalent circuit (b) of the improved CMOS bilateral switch*

switch to about 90R and virtually eliminates variations in its value with varying signal voltages, etc., as indicated by the equivalent circuit of *Figure 3.2(b)*. The only disadvantage of the *Figure 3.2* circuit is that it has a slightly lower leakage resistance than *Figure 3.1*.

Switch biasing

A CMOS bilateral switch can be used to switch or gate either digital or analogue signals, but must be correctly biased to suit the type of signal being controlled. *Figure 3.3* shows the basic ways of activating and biasing the bilateral switch. Figure 3.3(a) shows that the switch can be turned ON (closed) by taking the control terminal to V_{DD}, or turned OFF (open) by taking the control terminal to V_{SS}.

In digital signal switching applications (*Figure 3.3(b)*) the bilateral switch can be used with a single-ended supply, with V_{SS} to zero volts and V_{DD} at a positive value equal to (or greater than) that of the digital signal (up to a maximum of +15 volts). In analogue switching applications (*Figure 3.3(c)*) a split supply (either true or

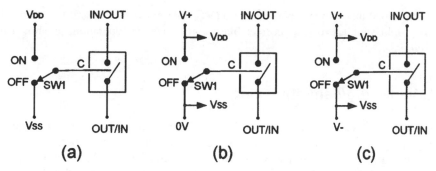

Figure 3.3 *Basic methods of turning the bilateral switch on and off* (a), *and power supply connections for use with* (b) *digital and* (c) *analogue IN/OUT signals.*

effective) must be used, so that the signal is held at a mean value of 'zero' volts; the positive supply rail goes to V_{DD}, which must be greater than the peak positive voltage value of the input signal, and the negative rail goes to V_{SS} and must be greater than the peak negative value of the input signal; the supply values are limited to plus or minus 7.5volts maximum. Typically, the bilateral switch introduces less than 0.5 percent of signal distortion when used in the analogue mode.

Logic level conversion

Note from the above description of the analogue system that, if a split supply is used, the switch control signal must switch to the positive rail to turn the bilateral switch on, and to the negative rail to turn the switch off. This arrangement is inconvenient in many practical applications. Consequently, some CMOS bilateral switch ICs (notably the 4051B to 4053B family) have built-in 'logic level conversion' circuitry which enables the bilateral switches to be controlled by a

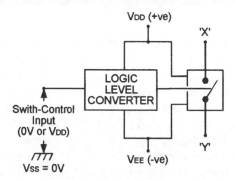

Figure 3.4 *Some ICs feature internal 'logic level conversion', enabling an analogue switch to be controlled via a single-ended input*

digital signal that switches between zero (V_{SS}) and positive (V_{DD}) volts, while still using split supplies to give correct biasing for analogue operation, as shown in *Figure 3.4*.

Multiplexing/demultiplexing

A multiplexer is a device that enables information from a number (*n*) of independent data lines to be individually selected and applied to a single data or output line, and acts like a single-pole multi-way electronic data selection switch. A demultiplexer is the opposite of a multiplexer, and enables the input from a single data line to be distributed to any selected one of a number (*n*) of independent output lines, and acts like a single-pole multi-way data distribution switch.

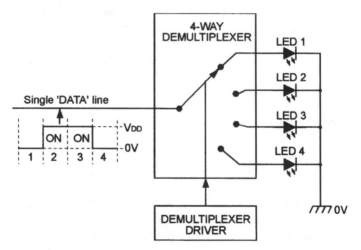

Figure 3.5   *A 4-way demultiplexer used to control four LEDs via a single DATA line*

Figure 3.5 shows how a 4-way demultiplexer (represented by a 4-way switch) can be used to control (turn on or off) four LEDs down a single DATA line. Assume here that the demultiplex driver continuously sequences the demultiplexer through the 1–2–3–4 cycle at a fairly rapid rate, and is synchronised to the 1–2–3–4 segments of the DATA line. Thus, in each cycle, in the '1' period LED 1 is off; in the '2' period LED 2 is on; in the '3' period LED 3 is on, and in the '4' period LED 4 is off. The state of each of the four LEDs is thus controlled via the logic bit of the single (sequentially time-shared) DATA line.

Figure 3.6 shows how a 4-way multiplexer can be used to feed three independent 'voice' signals down a single cable, and how a demultiplexer can be used to convert signals back into three independent voice signals at the other end. In practice, each

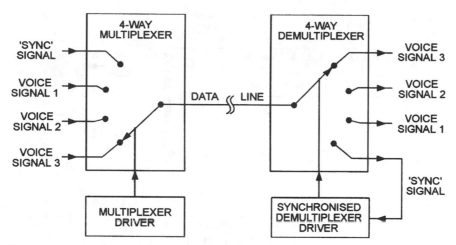

Figure 3.6 *A 4-way multiplexer/demultiplexer combination used to feed three independent voice signals through a single DATA line*

'sample' period of the DATA line must be short relative to the period of the highest voice frequency; the 'sync' signal is used to synchronise the timing at the two ends of the DATA line.

Note from the above description that a CMOS n-channel multiplexer can be regarded as a single-pole n-way bilateral switch, and that a CMOS multiplexer can be converted into a demultiplexer by simply transposing the notations of its input and output terminals. An n-way single-pole bilateral switch can thus be legitimately described as an n-channel multiplexer/demultiplexer. TTL ICs, on the other hand, are based on a unidirectional technology, and a dedicated TTL IC can be designed to act as either a multiplexer or demultiplexer, but (unlike a CMOS bilateral switch) cannot be designed to perform both functions.

Practical ICs

There are three major families of CMOS bilateral-switch ICs. The best known of these are the 4016B/4066B types, which are quad bilateral switches, each housing four independently accessible SPST bilateral switches, as already shown in Chapter 1 (in *Figure 1.27*). The 4016B uses the simple construction shown in *Figure 3.1*, and is recommended for use in sample-and-hold applications where low leakage impedance is of prime importance. The 4066B uses the improved type of construction of *Figure 3.2*, and is recommended for use in all applications where a low ON resistance is of prime importance.

The second family of ICs are the 4051B to 4053B types (*Figures 1.28, 3.7* and *3.8*). These are multi-channel multiplexer/demultiplexer ICs featuring built-in logic-level conversion. These ICs have three 'power supply' terminals (V_{DD}, V_{SS}, and

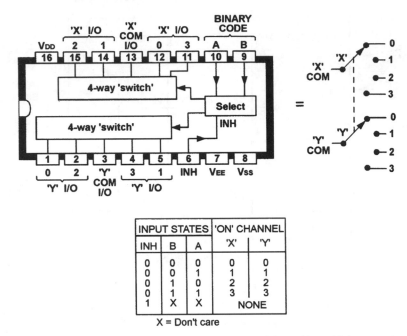

INPUT STATES			'ON' CHANNEL	
INH	B	A	'X'	'Y'
0	0	0	0	0
0	0	1	1	1
0	1	0	2	2
0	1	1	3	3
1	X	X	NONE	

X = Don't care

Figure 3.7   *The 4052B differential 4-channel multiplexer/demultiplexer acts as a ganged 2-pole 4-way switch*

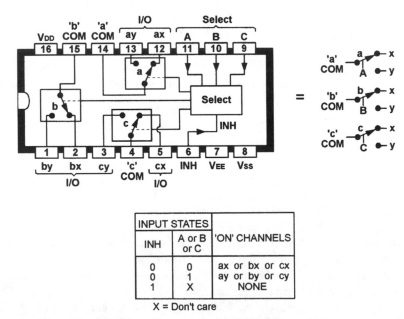

INPUT STATES		'ON' CHANNELS
INH	A or B or C	
0	0	ax or bx or cx
0	1	ay or by or cy
1	X	NONE

X = Don't care

Figure 3.8   *The 4053B triple 2-channel multiplexer/demultiplexer acts as three independent single-pole 2-way switches*

V_{EE}). In all applications, V_{DD} is taken to the positive supply rail, V_{SS} is grounded, and all digital control signals (for channel-select, inhibit, etc.) use these two terminals as their logic-reference values, i.e. logic-1 = V_{DD}, and logic-0 = V_{SS}. In digital signal processing applications, terminal V_{EE} is grounded (tied to V_{SS}). In analogue signal processing applications, V_{EE} must be taken to a -ve supply rail; ideally, V_{EE} = -(V_{DD}). In all cases, the V_{EE}-to-V_{DD} voltage must be limited to 15 volts maximum.

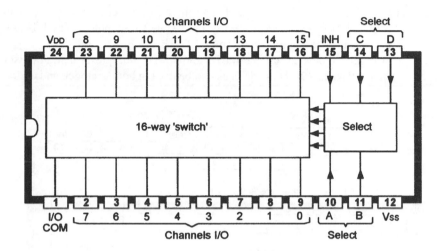

INPUT STATES				'ON'
INH	D C B A			CHANNEL
0	0	0 0 0		0
0	0	0 0 1		1
0	0	0 1 0		2
0	0	0 1 1		3
0	0	1 0 0		4
0	0	1 0 1		5
0	0	1 1 0		6
0	0	1 1 1		7
0	1	0 0 0		8
0	1	0 0 1		9
0	1	0 1 0		10
0	1	0 1 1		11
0	1	1 0 0		12
0	1	1 0 1		13
0	1	1 1 0		14
0	1	1 1 1		15
1	X	X X X		NONE

X = Don't care

Figure 3.9 *The 4067B 16-channel multiplexer/demultiplexer acts as a single-pole 16-way switch*

The 4051B (*Figure 1.28*, in Chapter 1) is an 8-channel multiplexer/demultiplexer, and can be regarded as a single-pole, 8-way bilateral switch. The IC has three binary control inputs (A, B and C) and an INHIBIT input. The three binary signals select one of the 8 channels to be turned on, as shown in the table.

The 4052B (*Figure 3.7*) is a differential 4-channel multiplexer/demultiplexer, and can be regarded as a ganged 2-pole, 4-way bilateral switch. It has two binary control inputs, which select the one of the four pairs of channels to be turned on, as shown in the table.

The 4053B (*Figure 3.8*) is a triple 2-channel multiplexer/demultiplexer, and can be regarded as a set of three independently accessible single-pole 2-way bilateral switches, each controlled via a single terminal (A, B or C).

The final family of devices are the 4067B and 4097B multiplexer/demultiplexer types (*Figures 3.9* and *3.10*). These devices can be used in both analogue and digital

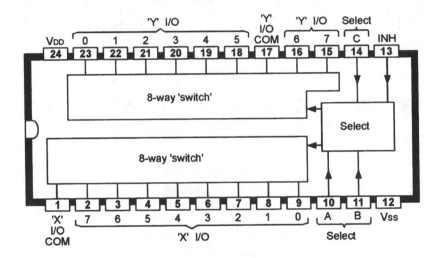

INPUT STATES				SELECTED CHANNELS	
INH	C	B	A	X	Y
0	0	0	0	0	0
0	0	0	1	1	1
0	0	1	0	2	2
0	0	1	1	3	3
0	1	0	0	4	4
0	1	0	1	5	5
0	1	1	0	6	6
0	1	1	1	7	7
1	X	X	X	NONE	

X = Don't care

Figure 3.10 *The 4097B differential 8-channel multiplexer/demultiplexer acts as a ganged 2-pole 8-way switch*

applications, but do not feature built-in logic-level conversion. The 4067B is a 16-channel device, and can be regarded as a single-pole 16-way bilateral switch. The 4097B is a differential 8-channel device, and can be regarded as a ganged 2-pole 8-way bilateral switch. Each IC is housed in a 24-pin DIL package.

Using 4016B/4066B ICs

The 4016B and 4066B are very versatile ICs, but a few simple precautions must be taken when using them, as follows:

(1) Input and switching signals must never be allowed to rise above V_{DD} or below V_{SS}.
(2) Each unused section of the IC must be disabled (see *Figure 3.11*) by taking its control terminal to V_{DD} and wiring one of its switch terminals to V_{DD} or V_{SS}, or by taking all three terminals to V_{SS}.

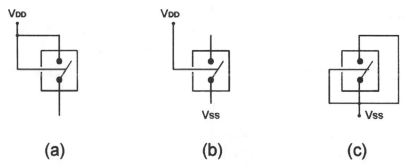

(a)   **(b)**   **(c)**

Figure 3.11   *Unused sections of the 4066B must be disabled, using any one of the connections shown here*

Figures 3.12 to *3.17* show some simple applications of the 4066B (or 4016B). *Figure 3.12* shows the device used to implement the four basic switching functions of SPST, SPDT, DPST and DPDT. *Figure 3.12(a)* shows the SPST connection, which has already been discussed. The SPDT function (*Figure 3.12(b)*) is implemented by wiring an inverter stage (a 4001B or 4011B, etc.) between the IC1a and IC1b control terminals. The DPST switch (*Figure 3.12(c)*) is simply two SPST switches sharing a common control terminal, and the DPDT switch (*Figure 3.12(d)*) is two SPDT switches sharing an inverter stage in the control line.

Note that the switching functions of *Figure 3.12* can be expanded or combined in any desired way by using more IC stages. Thus, a 10-pole 2-way switch can be made by using five of the *Figure 3.12(d)* circuits with their control lines tied together.

Each 4066B bilateral switch has a typical ON resistance of about 90R. *Figure 3.13* shows how four standard switch elements can be wired in parallel to make a single switch with a typical ON resistance of only 22.5 ohms.

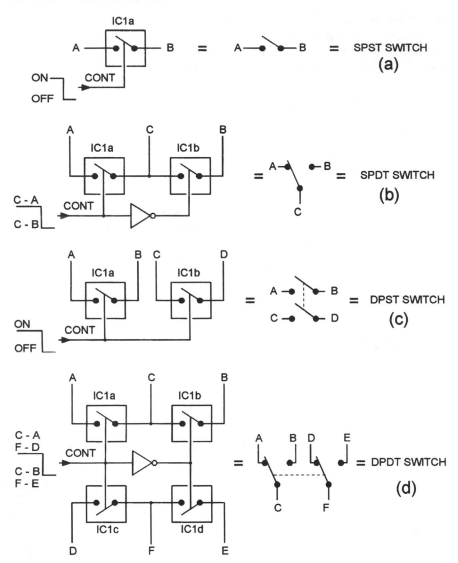

Figure 3.12 *Implementation of the four basic switching functions via the 4066B (IC1)*

Figures 3.14 to *3.17* show ways of using a bilateral switch as a self-latching device. In these circuits the switch current flows to ground via R3, and the control terminal is tied to the top of R3 via R2. Thus, in *Figure 3.14*, when PB1 is briefly closed the control terminal is pulled to the positive rail and the bilateral switch closes. With the bilateral switch closed, the top of R3 is at supply-line potential and, since the control terminal is tied to R3 via R2, the bilateral switch is thus latched on.

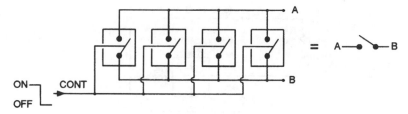

Figure 3.13 *This SPST switch has a typical ON resistance of only 22.5 ohms*

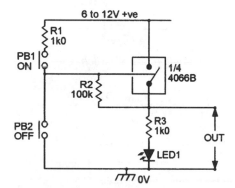

Figure 3.14 *Latching push-button switch*

Once latched, the switch can only be turned off again by briefly closing PB2, at which point the bilateral switch opens and the R3 voltage falls to zero. Note here that LED1 merely indicates the state of the bilateral switch, and R1 prevents supply line shorts if PB1 and PB2 are both closed at the same time.

Figure 3.15 shows how the above circuit can be made to operate as a latching touch-operated switch by increasing R2 to 10M and using R4–C1 as a 'hum' filter,

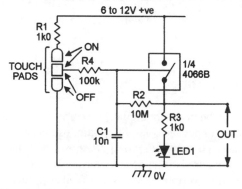

Figure 3.15 *Latching touch switch*

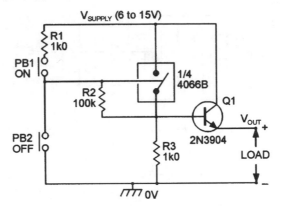

Figure 3.16   *Latching push-button power switch*

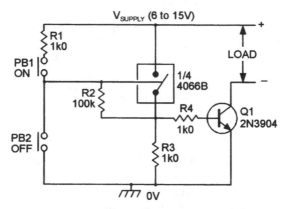

Figure 3.17   *Alternative version of the latching power switch*

and *Figures 3.16* and *3.17* show alternative ways of using the *Figure 3.14* circuit to connect power to external circuitry. The *Figure 3.16* circuit connects the power via a voltage follower stage, and the *Figure 3.17* design connects the power via a common-emitter amplifier.

Digital control circuits

Bilateral switches can be used to digitally control or vary effective values of resistance, capacitance, impedance, amplifier gain, oscillator frequency, etc., in any desired number of discrete steps. *Figure 3.18* shows how the four switches of a single 4066B can, by either shorting or not shorting individual resistors in a chain, be used to vary the effective total value of the resitance chain in sixteen digitally controlled steps of 10k each.

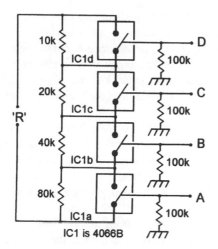

A	B	C	D	'R'
0	0	0	0	150k
0	0	0	1	140k
0	0	1	0	130k
0	0	1	1	120k
0	1	0	0	110k
0	1	0	1	100k
0	1	1	0	90k
0	1	1	1	80k
1	0	0	0	70k
1	0	0	1	60k
1	0	1	0	50k
1	0	1	1	40k
1	1	0	0	30k
1	1	0	1	20k
1	1	1	0	10k
1	1	1	1	0k

Figure 3.18 *16-step digital control of resistance. R can be varied from zero to 150k in 10k steps*

In practice, the step magnitudes of the *Figure 3.18* circuit can be given any desired value (determined by the value of the smallest resistor) so long as the four resistors are kept in the ratio 1–2–4–8. The number of steps can be increased by adding more resistor/switch stages; thus, a 6-stage circuit (with resistors in the ratio 1–2–4–8–16–32) will give resistance variation in sixty-four steps.

Figure 3.19 shows how four switches can be used to make a digitally controlled capacitor that can be varied in sixteen steps of 1n0 each. Again, the circuit can be expanded to give more 'steps' by simply adding more stages.

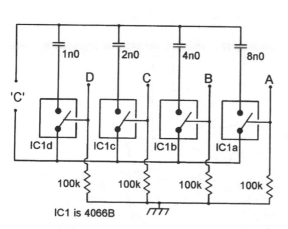

A	B	C	D	'C'
0	0	0	0	0
0	0	0	1	1n0
0	0	1	0	2n0
0	0	1	1	3n0
0	1	0	0	4n0
0	1	0	1	5n0
0	1	1	0	6n0
0	1	1	1	7n0
1	0	0	0	8n0
1	0	0	1	9n0
1	0	1	0	10n
1	0	1	1	11n
1	1	0	0	12n
1	1	0	1	13n
1	1	1	0	14n
1	1	1	1	15n

Figure 3.19 *16-step digital control of capacitance. C can be varied from zero to 15n in 1n0 steps*

Note in the *Figure 3.18* and *3.19* circuits that the resistor/capacitor values can be controlled by operating the 4066B switches manually, or automatically via simple logic networks, or via up/down counters, or via microprocessor control, etc.

The circuits of *Figure 3.18* and *3.19* can be combined in a variety of ways to make digitally controlled impedance and filter networks, etc. *Figure 3.20*, for example, shows two alternative ways of using them to make a digitally controlled 1st-order low-pass filter.

Digital control of amplifier gain can be obtained by hooking the *Figure 3.18* circuit into the feedback or input path of a standard op-amp inverter circuit, as shown in *Figures 3.21* and *3.22*. The gain of such a circuit equals R_f/R_{IN}, where R_f is the feedback resistance and R_{IN} is the input resistance. Thus, in *Figure 3.21*, the gain can be varied from zero to unity in sixteen steps of 1/15th each, giving a sequence of 0/15 (i.e. zero), 1/15, 2/15, etc., up to 14/15 and (finally) 15/15 (i.e. unity).

In the *Figure 3.22* circuit the gain can be varied from unity to ×16 in sixteen steps, giving a gain sequence of 1, 2, 3, 4, 5, etc. Note in both of these circuits that

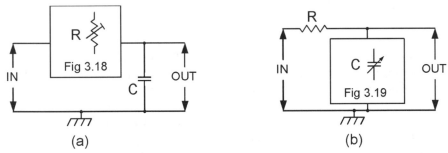

(a)   (b)

Figure 3.20 *Alternative ways of using* Figure 3.18 *or* 3.19 *to make a digitally controlled 1st-order low-pass filter*

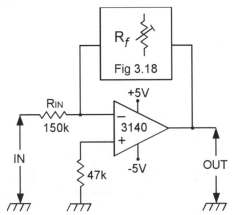

Figure 3.21 *Digital control of gain, using the* Figure 3.18 *circuit. Gain is variable from zero to unity in sixteen steps*

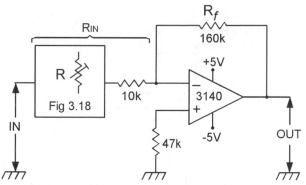

Figure 3.22   *Digital control of gain, using the* Figure 3.18 *circuit. Gain is variable from unity to* ×16 *in sixteen steps*

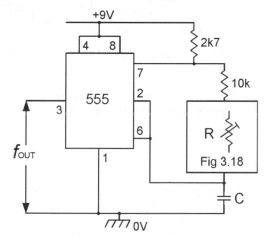

Figure 3.23   *Digital control of 555 astable frequency, in sixteen steps*

the op-amp uses a split power supply, so the 4066B control voltage must switch between the negative and positive supply rails.

Figure 3.23 shows how the *Figure 3.18* circuit can be used to vary the frequency of a 555 astable oscillator in sixteen discrete steps. Finally, *Figure 3.24* shows how three bilateral switches can be used to implement digital control of decade range selection of a 555 astable oscillator. Here, only one of the switches must be turned on at a time. Naturally, the circuits of *Figure 3.23* and *3.24* can easily be combined, to form a wide-range oscillator that can be digitally controlled via a microprocessor, etc.

Synthesised multigang pots, etc

One of the most useful applications of the bilateral switch is in synthesising multigang rheostats, pots, and variable capacitors in ac signal-processing circuitry.

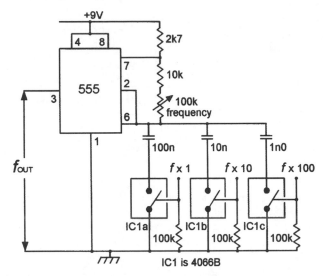

Figure 3.24 *Digital control of decade range selection of a 555 astable*

The synthesising principle is quite simple and is illustrated in *Figure 3.25*, which shows the circuit of a 4-gang 10k-to-100k rheostat for use at signal frequencies up to about 15kHz.

Here, the 555 is used to generate a 50kHz rectangular waveform that has its mark–space (M/S)-ratio variable from 11:1 to 1:11 via RV1, and this waveform is used to control the switching of the 4066B stages. All of the 4066B switches are fed with the same control waveform, and each switch is wired in series with a range resistor (Ra, Rb, etc.), to form one gang of the 'rheostat' between the a–a, b–b, etc., terminals.

Since the switching rate of this circuit is fast (50kHz) relative to the intended maximum signal frequency (15kHz), it can be seen that the *mean* or effective value

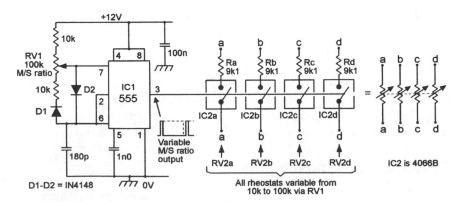

Figure 3.25 *Synthesised precision 4-gang 'rheostat'*

(when integrated over a few switching cycles) of each 'rheostat' resistance can be varied via M/S-ratio control RV1.

Thus, if IC2a is closed for 90 percent and open for 10 percent of each duty cycle (M/S ratio = 9:1), the apparent (mean) value of the a–a resistance will be 10 percent greater than Ra, i.e. 10k. If the duty cycle is reduced to 50 percent, the apparent Ra value doubles, to 18.2k. If the duty cycle is further decreased, so that IC2a is closed for only 10 percent of each duty cycle (M/S ratio = 1:9), the apparent value of Ra increase by a decade, to 91k. Thus, the apparent value of each gang of the 'rheostat' can be varied via RV1.

There are some important points to note about the above circuit. First, it can be given any desired number of 'gangs' by simply adding an appropriate number of switch stages and range resistors. Since all switches are controlled by the same M/S-ratio waveform, perfect tracking is automatically assured between the gangs. Individual gangs can be given different ranges, without affecting the tracking, by giving them different range resistor values. Also, the 'sweep' range and 'law' of the rheostats can be changed by simply altering the characteristics of the M/S-ratio generator. Finally, note that the switching control frequency of the circuit *must* be far higher than the maximum signal frequency that is to be handled, or the circuit will not perform correctly.

The 'rheostat' circuit of *Figure 3.25* can be made to function as a multi-gang variable capacitor by using ranging capacitors in place of ranging resistors. In this case, however, the apparent capacitance value decreases as the duty cycle is decreased.

The *Figure 3.25* principle can be expanded, to make synthesised multi gang pots, by using the basic technique shown in *Figure 3.26*. Here, in each gang, two 'rheostats' are wired in series but have their switch control signals fed in anti-phase,

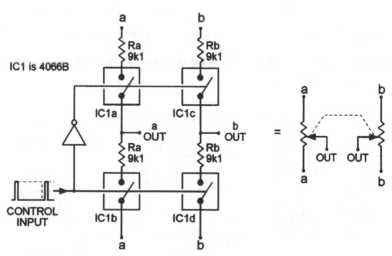

Figure 3.26 *Synthesised precision 2-gang 'potentiometer'*

so that one rheostat value increases as the other decreases, thus giving a variable potential-divider action. This basic circuit can be expanded, to incorporate any desired number of gangs, by simply adding more double-rheostat stages.

Miscellaneous applications

To complete this look at the CMOS bilateral switch, *Figures 3.27* and *3.28* show a couple of miscellaneous fast-switching applications of the device. *Figure 3.27* shows how it can be used as a sample-and-hold element. The 3140 CMOS op-amp is used as a voltage follower and has a near-infinite input impedance, and the 4016B switch also has a near-infinite impedance when open. Thus, when the 4016B is closed, the 10n capacitor rapidly follows all variations in input voltage, but when the switch opens the prevailing capacitance charge is stored and the resulting voltage remains available at the op-amp output.

Finally, *Figure 3.28* shows a bilateral switch used in a linear ramp generator. Here, the op-amp is used as an integrator, with its non-inverting pin biased at 5 volts

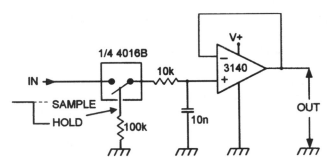

Figure 3.27 *Using a bilateral switch as a sample-and-hold element*

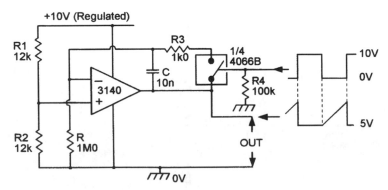

Figure 3.28 *Using the 4066B to implement a ramp generator circuit*

via R1–R2, so that a constant current of 5µA flows into the inverting pin via R. When the bilateral switch is open, this current linearly charges capacitor C, causing a rising ramp to appear at the op-amp output. When the bilateral switch closes, C is rapidly discharged via R3 and the output switches down to 5 volts.

The basic *Figure 3.28* circuit is quite versatile. The switching can be activated automatically via a free-running oscillator or via a voltage-trigger circuit. Bias levels can be shifted by changing the R1–R2 values or can be switched automatically via another 4066B stage, enabling a variety of ramp waveforms to be generated.

4

AC power control circuits

The easiest way of electronically controlling the power feed to lamps, heaters, motors, or other devices that are operated directly from the standard 50Hz to 60Hz single-phase AC power line is via solid-state devices such as rectifiers, SCRs, or triacs. The basic operating theory and application principles of simple SCRs and triacs have already been described in Chapter 1; most of the practical circuits shown in the present chapter are designed around these devices, and can be used with either 115 volt or 230 volt AC power lines (where applicable, 115 volt AC component values are shown in parenthesis in the circuit diagrams); in each of these circuits, the user must simply select the triac or SCR rating to suit his/her own particular application. Before looking at practical SCR and triac circuits, however, the chapter starts off by looking at some half-wave rectifier circuits that can be used to give very simple AC power control of lamps, motors, and soldering irons.

Half-wave rectifier circuits

The simplest type of 'half-wave rectifier' AC power control circuit is that shown in *Figure 4.1*, in which the AC line voltage (V_{IN}) is specified in volts r.m.s. (usually 230V or 115V). Here, zero power is applied to the load when S1 is switched to its '1' (OFF) position, and full power is fed to the load when S1 is in its '3' (ON) position. When S1 is in its '2' (PART) position, the relative power reaching the load depends on the nature of the load. If the load is purely capacitive, the circuit acts like a peak voltage detector, and the load voltage equals 1.41 $\times V_{IN}$, but if the load is purely resistive, the r.m.s load voltage equals 0.707 $\times$ V_{IN} (see Chapter 9) and – since power is proportional to the square of the applied r.m.s. voltage – the load dissipates one-half of maximum power.

In practice, very few loads are purely resistive, and *Figures 4.2* to *4.4* show how the basic half-wave rectifier circuit can be adapted to give two-level power control of lamps, electric drills, and soldering irons that are operated from the AC power lines. Note in each of these circuits that the r.m.s. voltage fed to the load equals V_{IN} when S1 is in position 3, or 0.707 $\times$ V_{IN} when S1 is in position 2.

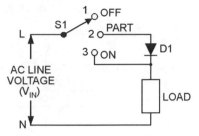

Figure 4.1   *Basic '2-level' AC power control circuit*

The *Figure 4.2* circuit uses a lamp load, which has a resistance that rises with filament temperature; when operated at 0.707 times maximum voltage the lamp resistance is about three-quarters of maximum, causing the lamp to generate roughly half of maximum brilliance when S1 is switched to the DIM position.

The *Figure 4.3* circuit uses the universal (AC/DC) motor of an electric drill, for example, as its load. Such motors have an inherent self-regulating speed-control capacity, and because of this the motor operates (when lightly loaded) at about 70 percent of maximum speed when S1 is in the PART position.

The *Figure 4.4* circuit uses a soldering iron heating element as its load, and these have a resistance that increases moderately with temperature; thus, when the iron is operated at '0.707' voltage its resistance is slightly reduced, the net effect being that

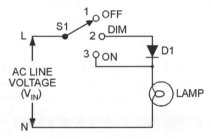

Figure 4.2   *Lamp burns at half brilliance in DIM position*

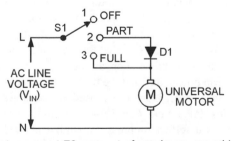

Figure 4.3   *Drill motor runs at 70 percent of maximum speed in PART position*

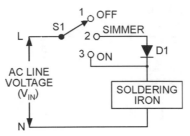

Figure 4.4 *Soldering iron operates at two-thirds power in SIMMER position*

the iron operates at about two-thirds of maximum power when S1 is in the SIMMER position, thus keeping the iron hot but not to such a degree that its bit deteriorates.

Note in the *Figure 4.1* to *4.4* circuits that D1 can be any type that has a current rating greater than the load requirement and has a peak reverse-voltage rating of at least $1.41 \times V_{IN}$.

Triac power switches

Triacs are, as already explained in Chapter 1, solid-state power switches that can be triggered (turned on and latched) either synchronously or non-synchronously with the AC power-line voltage, but which automatically turn off at the end of each power-line half-cycle as their main-terminal currents fall below the device's 'minimum holding' value.

When used as simple power switches, synchronous circuits *always* turn on at the same point in each AC half-cycle (usually just after the zero-crossing point) and can generate minimal RFI. The trigger points of non-synchronous circuits, on the other hand, are not always synchronised to a fixed point of the AC cycle, and may generate significant RFI, particularly at the point of initial turn-on.

Figures 4.5 to *4.12* show a variety of non-synchronous triac power switch circuits that can be used in basic ON/OFF line-switching applications. The basic action of the *Figure 4.5* circuit was explained in Chapter 1, and is such that the triac is off and acts like an open switch when SW1 is open, but acts like a closed switch that is gated on from the AC power line via the load and R1 shortly after the start of each AC half-cycle when SW1 is closed. Note that the triac's main terminal voltage drops to only a few hundred millivolts as soon as the triac turns on, so R1 and SW1 consume very little mean power, and that the triac's trigger point is *not* synchronised to the AC power-line when SW1 is initially closed, but becomes synchronised on all subsequent half-cycles. Also note that R2–C1 form a 'snubber' network that provides the triac with rate-effect suppression; similar networks are fitted to most triac circuits shown in this chapter.

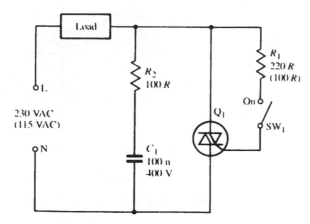

Figure 4.5 *Simple AC power switch, AC line triggered*

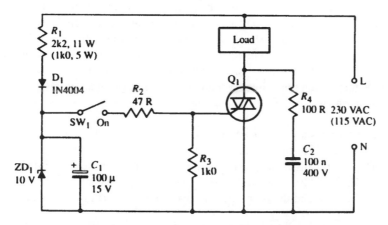

Figure 4.6 *AC power switch with line-derived DC triggering*

Figure 4.6 shows how the triac can be used as a power switch that can be triggered via a power-line-derived DC supply. C1 is charged to +10V on each positive power-line half-cycle via R1–D1, and the C1 charge triggers the triac when SW1 is closed. Note that R1 is subjected to almost the full AC line voltage at all times, and thus needs a fairly high power rating, and that all parts of this circuit are 'live', making it difficult to interface to external control circuitry.

Opto-coupler input control

Figure 4.7 shows how the above circuit can be modified so that it can easily be interfaced to external control circuitry via an optocoupler (the basic operating

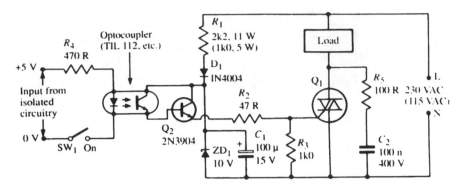

Figure 4.7   *Isolated-input (optocoupled) AC power switch, DC triggered*

principle of the opto-coupler is described in Chapter 1). Here, SW1 is simply replaced by transistor Q2, which in turn is driven from the photo-transistor side of a normal optocoupler. The LED side of the optocoupler is driven from a 5 volt (or greater) DC supply via R4. The triac turns on only when the external supply is connected via SW1. Optocouplers have insulation potentials as high as several thousand volts, so the 'external' part of this circuit is electrically fully isolated from the power-line-driven triac circuitry, and can easily be designed to give any desired form of automatic 'remote' triac operation by replacing SW1 with a suitable electronic switch.

Figure 4.8 shows a variation of the above circuit. Here, the triac is AC triggered in each half-cycle via the AC impedance of C1–R1 and via back-to-back zeners ZD1–ZD2, but C1 dissipates near-zero power. Bridge rectifier D1-to-D4 is wired across the ZD1–ZD2–R2 network and is loaded by Q2; when Q2 is off, the bridge is effectively open and the triac turns on just after the start of each AC half-cycle,

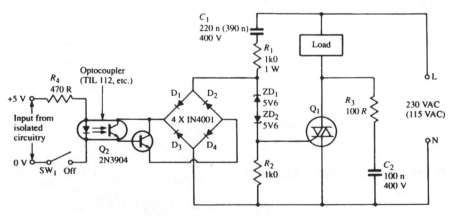

Figure 4.8   *Isolated-input AC power switch, AC triggered*

but when Q2 is on, a near-short appears across ZD1–ZD2–R2, so the triac is off. Q2 is actually driven via the optocoupler from the isolated external circuit, so the triac is normally on but turns off when SW1 is closed.

DC triggering

Figures 4.9 and *4.10* show ways of triggering a triac power switch via a transformer-derived DC supply and a transistor-aided switch. In *Figure 4.9* the transistor and the triac are both driven on when SW1 is closed, and are off when SW1 is open. In practice, SW1 can easily be replaced by an electronic switch, enabling the triac to be operated by heat, light, sound, time, etc. Note, however, that the whole of this circuit is 'live'; *Figure 4.10* shows the circuit modified for optocoupler operation, enabling it to be activated via fully isolated external circuitry.

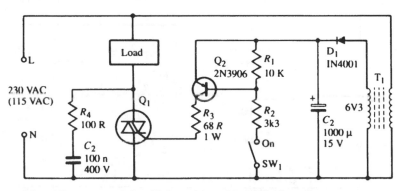

Figure 4.9 *AC power switch with transistor-aided DC triggering*

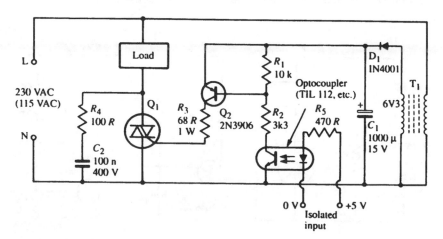

Figure 4.10 *Isolated-input AC power switch with DC triggering*

UJT triggering

Another way of driving a non-synchronous triac ON/OFF power switch is via a UJT (Unijunction transistor), and *Figures 4.11* and *4.12* show two ways of obtaining such triggering via a fully isolated external circuit; the triggering action is obtained via UJT oscillator Q2, which operates at several kHz and (as described in Chapter 1) feeds output pulses to the triac gate via special-purpose pulse transformer T1, which provides the desired 'isolation'. Because of its fairly high oscillating frequency, the UJT triggers the triac within a few degrees of the start of each power-line half-cycle when the oscillator is active.

In *Figure 4.11*, Q3 is wired in series with the UJT's main timing resistor, so the UJT and triac turn on only when SW1 is closed. In *Figure 4.12*, Q3 is wired in parallel with the UJT's main timing capacitor, so the UJT and triac turn on only

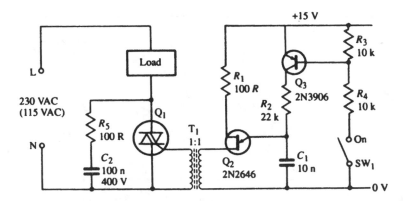

Figure 4.11   *Isolated-input (transformer-coupled) AC power switch*

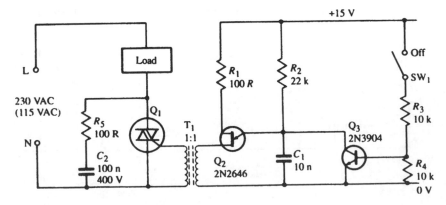

Figure 4.12   *Isolated-input AC power switch*

when SW1 is open. In each case, SW1 can easily be replaced by some type of electronic switch, to give automatic power switching in response to variations in time, light level, or temperature.

Optocoupled triacs

Triacs, like most semiconductor devices, are inherently photosensitive, and are available in both 'conventional' and 'optocoupler' forms. An optocoupled triac is made by mounting a photosensitive triac and a matching LED in a single package, in such a way that the two devices are optically coupled but are electrically fully isolated. Optocoupled triacs usually have very limited output current ratings (typically about 100mA r.m.s. at 400V peak), and are usually housed in a simple 6-pin DIL package such as that shown in *Figure 4.13*. The triac can (typically) handle brief surge currents of up to 1.2A for 10μs, or 400mA for 40μs.

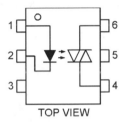

TOP VIEW

Figure 4.13 *Typical optocoupled triac*

Figures 4.14 to *4.16* show some practical ways of using an optocoupled triac with a 230V AC power supply; in all cases, R1 should be chosen to pass a LED 'ON' current in the range 20mA to 40mA. In *Figure 4.14*, the triac is used to directly activate a low-power filament lamp (less than 23W), which must have an r.m.s. rating of less than 100mA.

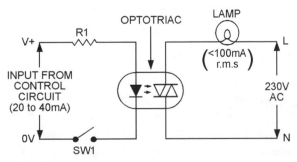

Figure 4.14 *Low-power lamp control*

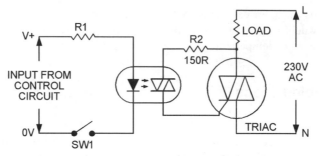

Figure 4.15 *High-power control via a triac slave*

In practice, an optocoupled triac is best used to activate a slave triac, as shown in *Figure 4.15*, which in turn can activate a load of any desired power rating. This circuit is suitable for use only with non-inductive loads such as lamps or heating elements. *Figure 4.16* shows how the circuit can be modified for use with inductive loads such as electric motors; the R2–C1–R3 network provides a degree of phase-shift to the triac gate-drive network, to ensure correct triac triggering action, and R4–C2 form a snubber network, to suppress 'rate' effects.

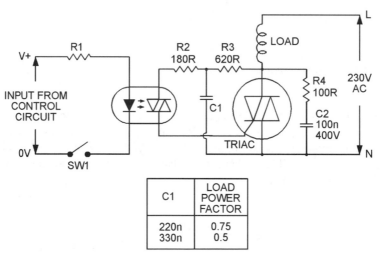

C1	LOAD POWER FACTOR
220n	0.75
330n	0.5

Figure 4.16 *Driving an inductive load*

Automatic control

The main advantage of the *Figure 4.7* to *4.16* triac circuits, when compared to ordinary electromechanical switching circuits, is that they can easily be modified to give automatic switching action in response to variations in time, light, or heat, etc., by simply using suitable circuitry in their input control position. An almost infinite variety of such control circuits can be devised, and many of the relay-output circuits of Chapter 2 can easily be modified to give direct triac activation.

Synchronous power switching

Synchronous AC power switches can easily be designed to generate minimal RFI, and *Figures 4.17* to *4.26* show a variety of synchronous triac power switch circuits that can be used in various ON/OFF line switching applications.

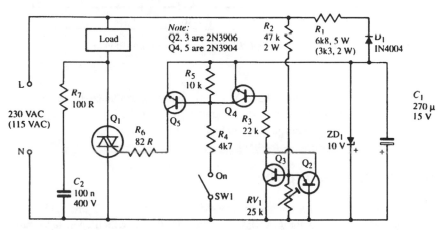

Figure 4.17 *'Transistorised' synchronous line switch*

Figure 4.17 shows a 'transistorised' synchronised AC power switch that is triggered near the zero-voltage cross-over points of the AC waveform. The triac gate trigger current is obtained from a 10V DC supply derived from the AC power-line via R1–D1–ZD1 and C1, and this supply is switched to the gate via Q5, which in turn is controlled by SW1 and zero-crossing detector Q2–Q3–Q4. The action of Q5 is such that it can turn on and conduct gate current when SW1 is closed and Q4 is off. The action of the zero-crossing detector is such that Q2 or Q3 are driven on whenever the instantaneous AC voltage is positive or negative by more than a few volts (depending on the setting of RV1), thereby driving Q4 on via R3 and inhibiting Q5. Thus, gate current can only be fed to the triac when SW1 is closed and the instantaneous AC power-line voltage is within a few volts of zero; this ON/OFF switching circuit thus generates minimal switching RFI.

Figure 4.18 shows how the above circuit can be modified so that the triac can only turn on when SW1 is open. Note in both circuits that, since only a narrow pulse of gate current is sent to the triac, the *mean* consumption of the DC supply is very

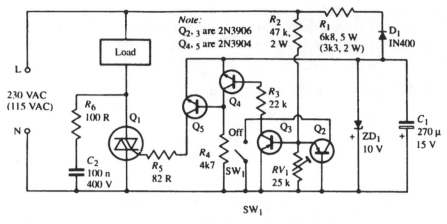

Figure 4.18   *Alternative version of the 'transistorised' line switch*

low (about 1mA). Also note that SW1 can easily be replaced by an electronic switch, to give automatic operation via heat, light, time, etc., or by an optocoupler to give fully isolated operation from external circuitry.

Special triac-gating ICs

Several dedicated synchronous zero-cross-over triac-gating ICs are available; the best known of these are the CA3059 and the TDA1024, which each have built-in power-line-derived DC power supply circuitry, a zero-crossing detector, triac gate-drive circuitry, and a high-gain differential amplifier/gating network. *Figure 4.19* shows the internal circuitry of the CA3059, together with its minimal external connections. AC line power is connected to pins 5 and 7 via limiting resistor R_s (22k, 5W when 230V AC is used). D1 and D2 act as back-to-back zeners and limit the pin-5 voltage to ±8V. On positive half-cycles D7 and D13 rectify this voltage and generate 6.5V across the 100μF pin-2 capacitor, which stores enough energy to drive all internal circuitry and provides adequate triac gate drive, with a few mA of spare drive available for powering external circuitry if needed.

Bridge rectifier D3 to D6 and transistor Q1 act as a zero-crossing detector, with Q1 being driven to saturation whenever the pin-5 voltage exceeds ±3V. Gate drive to an external triac can be made via the emitter (pin-4) of the Q8–Q9 Darlington pair, but is available only when Q7 is turned off. When Q1 is turned on (pin-5 greater than ±3V) Q6 turns off through lack of base drive, so Q7 is driven to saturation via R7 and no triac gate drive is available at pin-4. Triac gate drive is thus available only when pin-5 is close to the 'zero-voltage' AC power-line value, and is delivered in the form of a narrow pulse centred on the cross-over point; its power is supplied via C1.

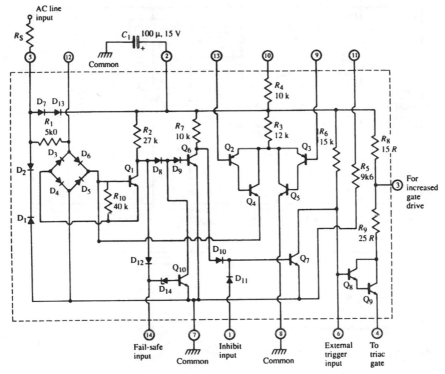

Figure 4.19 *Internal circuit and minimal external connections of the CA3059 synchronous 'zero-voltage' triac driver IC*

The CA3059 incorporates a general-purpose voltage comparator or differential amplifier, built around Q2 to Q5. Resistors R4 and R5 are externally available for biasing one side of the amplifier. The emitter current of Q4 flows via the base of Q1 and can be used to disable the triac gate drive (pin-4) by turning Q1 on. The configuration is such that the gate drive can be disabled by making pin-9 positive relative to pin-13. The drive can also be disabled by connecting external signals to pin-1 and/or pin-14.

Figures 4.20 and *4.21* show how the CA3059 can be used to give manually controlled 'zero voltage' ON/OFF switching of the triac. These two circuits use SW1 to enable or disable the triac gate drive via the IC's internal differential amplifier (the drive is enabled only when pin-13 is biased above pin-9). In the *Figure 4.20* circuit pin-9 is biased at half-supply volts and pin 13 is biased via R2–R3 and SW1, and the triac thus turns on only when SW1 is closed.

In *Figure 4.21*, pin-13 is biased at half-supply volts and pin 9 is biased via R2–R3 and SW1, and the triac again turns on only when SW1 is closed. In both circuits, SW1 handles maximum potentials of 6V and maximum currents of about 1mA.

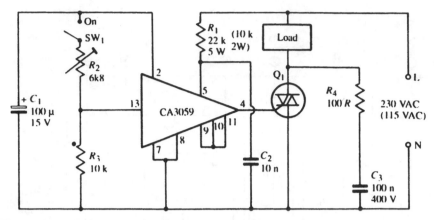

Figure 4.20   *Direct-switching IC-gated 'zero-voltage' line switch*

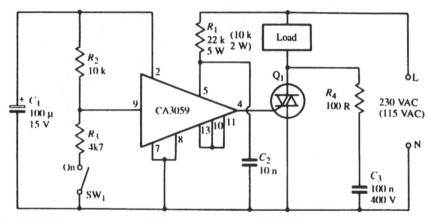

Figure 4.21   *An alternative method of direct switching the CA3059 IC*

Note in these designs that capacitor C2 is used to apply a slight phase delay to the pin-5 'zero-voltage-detecting' terminal, and causes the gate pulses to occur after (rather than to 'straddle') the zero-voltage point.

Note in the *Figure 4.21* circuit that the triac can be turned on by pulling R3 low or can be turned off by letting R3 float. *Figures 4.22* and *4.23* show how this simple fact can be put to use to extend the versatility of the basic circuit. In *Figure 4.22* the triac can be turned on and off by transistor Q2, which in turn can be activated by on-board CMOS circuitry (such as one-shots, astables, etc.) that are powered from the 6V pin-2 supply. In *Figure 4.23* the circuit can be turned on and off by fully isolated external circuitry via an inexpensive optocoupler, which needs an input in excess of only a couple of volts to turn the triac on.

Alternatively, *Figure 4.24* shows how the TDA1024 IC can be used in place of the CA3059 to give either directly switched or optocoupled 'zero-voltage' triac control.

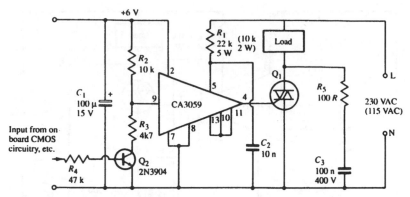

Figure 4.22 *Method of transistor switching the CA3059 via on-board CMOS circuitry, etc*

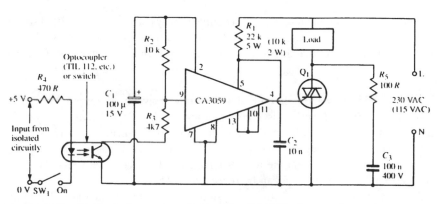

Figure 4.23 *Method of remote-switching the CA3059 via an optocoupler*

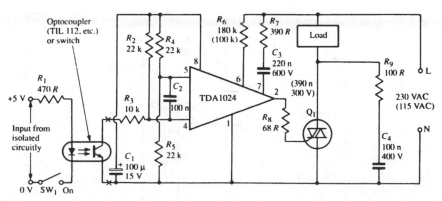

Figure 4.24 *The TDA1024 used to give either directly switched or optocoupled 'zero-voltage' triac control*

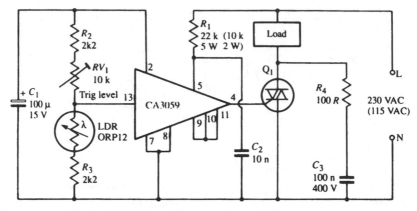

Figure 4.25   *Basic 'dark-activated' zero-voltage switch*

Finally, *Figures 4.25* and *4.26* show a couple of ways of using the CA3059 so that the triac operates as a light-sensitive 'dark-operated' power switch. In these designs the IC's built-in differential amplifier is used as a precision voltage comparator that turns the triac on or off when one of the comparator input voltages goes above or below the other.

Figure 4.25 shows a simple dark-activated power switch. Here, pin-9 is tied to half-supply volts and pin 13 is controlled via the R2–RV1–LDR–R3 potential divider. Under bright conditions the LDR has a low resistance, so pin-13 is below pin-9 and the triac is disabled. Under dark conditions the LDR has a high resistance, so pin-13 is above pin-9 and the triac is enabled and power is fed to the AC load. The threshold switching level can be preset via RV1.

Figure 4.26 shows how a degree of hysteresis or 'backlash' can be added to the above circuit, so that the triac does not switch annoyingly in response to small changes

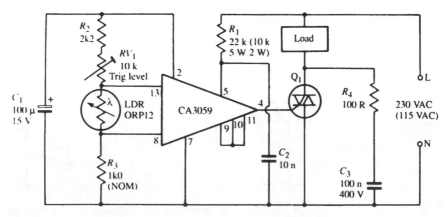

Figure 4.26   *Dark-activated zero-voltage switch with hysteresis provided via R_3*

(such as caused by passing shadows, etc.) in ambient light level. The hysteresis level is controlled via R3, which can be selected to suit particular applications.

Electric heater controllers

Triacs can be used to give automatic room temperature control by using electric heaters as the triac loads, and thermostats or thermistors as the triac feedback control elements. Two basic methods of heater control can be used, either simple automatic ON/OFF power switching, or fully automatic 'burst-fire' proportional power control, as described in Chapter 1. In the former case, the heater switches fully on when the room temperature falls below a pre-set level and turns off when it rises above the pre-set level. In the latter (burst-fire) case, the *mean* heater power is automatically adjusted so that, when the room temperature is at the precise pre-set level, the heater output power self-adjusts to exactly balance the thermal losses of the room.

Because of the high-power requirements of electric heaters, special care must be taken in the design of triac controllers to keep RFI generation to minimal levels. Two options are open to designers, to use either continuous DC gating of the triac, or to use synchronously pulsed gating. The advantage of DC gating is that, in basic ON/OFF switching applications, the triac generates zero RFI under normal (ON) running conditions; the disadvantage is that the triac may generate a very powerful RFI spike as it is initially switched on in its first ON half-cycle. The advantage of synchronous gating is that no high-level RFI is generated as the triac transitions from the OFF to the ON condition; the disadvantage is that the triac generates continuous very low level RFI under normal (ON) running conditions.

DC-gated circuits

Figures 4.27 and *4.28* show examples of DC-gated heater controller circuits, in which the DC supply is derived via T1–D1 and C1, and the heater can be controlled

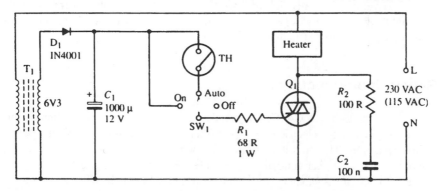

Figure 4.27 *Heater controller with thermostat-switched DC gating*

either manually or automatically via SW1. The *Figure 4.27* circuit is auto-controlled via a thermostat, and calls for no further explanation.

The *Figure 4.28* circuit, on the other hand, is controlled by n.t.c. (negative-temperature coefficient) thermistor TH1 and transistors Q2–Q3, and calls for some explanation. RV1–TH1–R2–R3 are used as a thermal bridge, with Q2 acting as a bridge-balance detector. RV1 is adjusted so that Q2 just starts to turn on as the temperature falls to the desired pre-set level; below this level, Q2–Q3 and the triac are all driven hard on, and above this level all three components are cut off.

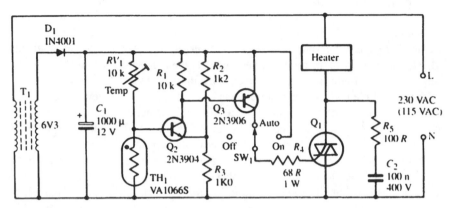

Figure 4.28   *Heater controller with thermistor-switched DC gating*

Note in this circuit that, since the gate-drive polarity is always positive but the triac main-terminal current is alternating, the triac is gated alternately in the so-called I+ and III+ modes or quadrants, and that the gate sensitivities are quite different in these two modes. Consequently, when the TH1 temperature is well below the pre-set level Q3 is driven hard on and the triac is gated in both quadrants and gives full power drive to the heater, but when the temperature is very close to the pre-set value Q3 is only lightly driven on, so the triac is gated in the I+ mode only and the heater operates at only half of maximum power drive. The circuit thus gives very fine control of temperature.

Synchronous circuits

Figure 4.29 shows how a CA3059 IC can be used to make an automatic thermistor-regulated synchronous electric heater controller that gives a simple ON/OFF heater switching action. The circuit is similar to the 'dark-activated' power switch of *Figure 4.25* except that n.t.c thermistor TH1 is used as the feedback sensing element; the circuit is capable of maintaining room temperature within a degree or so of the value pre-set via RV1.

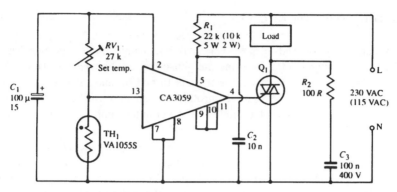

Figure 4.29 *Heater controller with thermistor-regulated zero-voltage switching*

Finally, to complete this 'heater controller' section, *Figure 4.30* shows a 'burst-fire' or proportional heater controller that is capable of regulating room temperatures to within ±0.5°C of a pre-set value. Here, a thermistor-controlled voltage is applied to the pin-13 side of the CA3059's comparator, and a repetitive 300ms ramp waveform, centred on half-supply volts, is applied to the pin-9 side of the comparator from CMOS astable IC1. The circuit action is such that the triac is synchronously gated fully on if the ambient temperature is more than a couple of degrees below the pre-set level, or is cut off if it is more than a couple of degrees above the pre-set level, but when it is close to the pre-set value the ramp waveform comes into effect and synchronously turns the triac on and off in the burst-fire or integral cycle mode once every 300ms, with a mark–space (M/S) ratio proportional to the temperature differential.

Thus, if the M/S ratio is 1:1, the heater generates only half of maximum power, and if the ratio is 1:3 it generates only one-quarter of maximum power. The net effect is that the heater output power self-adjusts to meet the room's heating needs. When the room temperature reaches the precise pre-set value, the heater does not switch fully off, but generates just enough output power to precisely match the thermal losses of the room. The system thus gives very precise temperature control.

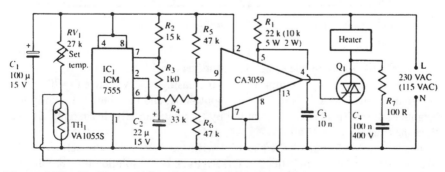

Figure 4.30 *Heater controller giving integral-cycle precision temperature regulation*

AC lamp dimmer circuits

Triacs can easily be used to make very efficient lamp dimmers (which vary the brilliance of filament lamps) by using the phase-triggered power control principles described in Chapter 1, in which the triac is turned on and off once in each AC power-line half-cycle, its M/S ratio controlling the mean power fed to the lamp. All such circuits require the use of a simple L–C filter in the lamp feed line, to minimise RFI problems.

The three most popular ways of obtaining variable phase-delay triac triggering are to use either a diac plus R-C phase-delay network, or to use a line-synchronised variable-delay UJT trigger, or to use a special-purpose IC as the triac trigger. *Figures 4.31* to *4.39* show practical examples of lamp dimmers using each of these three different methods.

Figure 4.31 shows a practical diac-triggered lamp dimmer, in which R1–RV1–C1 provide the variable phase delay. This circuit is similar to the basic lamp dimmer circuit described in Chapter 1, except for the addition of ON/OFF switch SW1, which is ganged to RV1 and enables the lamp to be turned fully off.

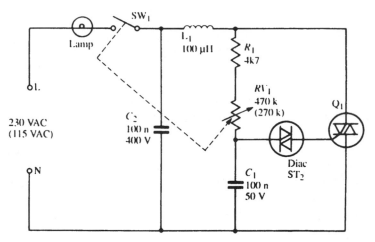

Figure 4.31   *Practical circuit of a simple lamp dimmer*

A weakness of this simple design is that it has considerable control hysteresis or backlash, e.g. if the lamp is dimmed off by increasing the RV1 value to 470k, it will not go on again until RV1 is reduced to about 400k, and then burns at a fairly high brightness level. This backlash is caused by the diac partially discharging C1 each time the triac fires.

The backlash effect can be reduced by wiring a 47R resistor in series with the diac, to reduce its discharge effect on C1. An even better solution is to use the gate slaving circuit of *Figure 4.32*, in which the diac is triggered from C2, which

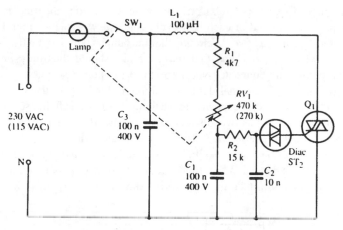

Figure 4.32 *Improved lamp dimmer with gate slaving*

'follows' the C1 phase-delay voltage but protects C1 from discharging when the diac fires.

If absolutely zero backlash is needed, the UJT-triggered circuit of *Figure 4.33* can be used. The UJT is powered from a 12V DC supply derived from the AC line via R1–D1–ZD1–C1 and is line-synchronised via the Q2–Q3–Q4 zero-crossing detector network, the action being such that Q4 is turned on (applying power to the UJT) at all times other than when the AC power-line voltage is close to the zero-cross-over point at the end and start of each AC half-cycle. Thus, shortly after the start of each half-cycle, power is applied to the UJT circuit via Q4, and some time

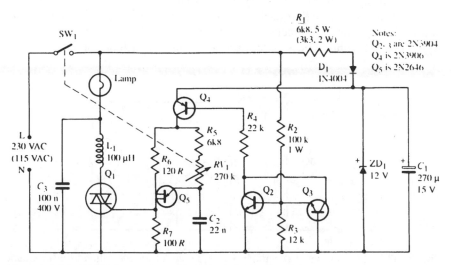

Figure 4.33 *UJT-triggered zero-backlash lamp dimmer*

later (determined by R5–RV1–C2) a trigger pulse is applied to the triac gate via Q5. The UJT resets at the end of each half-cycle, and a new sequence then begins.

Note in the *Figure 4.31* to *4.33* circuits that the lamp is wired to the live or 'phase' side of the AC power line, and that these circuits should thus only be used for driving lamps that are connected via a normal 3-pin power socket, and cannot be directly used with conventional domestic 'lighting' circuits. In the UK (and most other countries), the electric power to each floor of a small house or office is conventionally applied via two distinct wiring networks; the connections to the 3-pin power sockets are made via a 'ring' circuit and/or wiring 'spurs' (see Chapter 9), and the connections to all ceiling-mounted lamps (and most wall-mounted lamps) are made via dedicated 'lighting' circuits. National safety regulations require these lighting circuits to take the basic form shown in *Figure 4.34*, with (in each

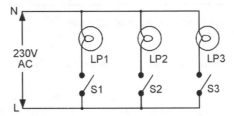

Figure 4.34   *Normal domestic lighting circuit*

lamp/switch section) one of the lamp-socket wires going directly to the neutral or 'safe' side of the power line, and with one side of its ON/OFF switch connected to the live or 'phase' line, thus ensuring that the entire lamp and socket are safely at 'neutral' potential when the power switch is open.

In practice, the *Figure 4.32* (and *4.31*) circuit can easily be modified for use with normal domestic lighting circuits, as shown in *Figure 4.35*. The *Figure 4.33* circuit would need a complete redesign to make it suitable for such operation.

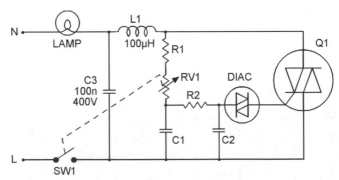

Figure 4.35   *Modified version of the* Figure 4.32 *circuit, for use with normal domestic lighting*

A 'smart' lamp dimmer

Many modern lamp dimmers have their triac driven via a dedicated 'smart' IC that takes its commands via a touch-sensitive pad or push-button input switch. Siemens are leaders in this field, and their most popular current-production IC (introduced in 1990) is the SLB0586 (see *Figure 4.36*), which needs a 5.6V, 0.45mA DC supply and incorporates touch-conditioning circuitry and an 'options' control (pin-2) that allows the user to select a variety of dimming or switching modes.

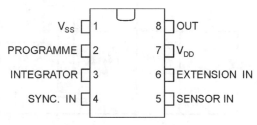

Figure 4.36 *SLB0586 outline and pin notations*

Figure 4.37 shows the SLB0586 basic application circuit, which is suitable for use in conventional 'lighting' networks and uses a single touch-sensitive input control. The IC is powered (between pins 1 and 7) from a 5.6V DC supply derived from the AC line via R2–C2–ZD1–D1–C3, and has its EXTENSION input disabled by shorting pin-6 to pin-7. The pin-5 SENSOR input works on the induced pick-up principle, in which the operator's body picks up radiated AC power-line signals that are detected via the conductive touch pad, which must be placed close to the IC to avoid false pick-up. The operator is protected from the power-line voltage via R7–R8; for correct operation the AC power lines must be connected as shown, with the live or 'hot' lead to pin-1 of the IC, and the neutral line to the lamp.

The SLB0586 has three basic operating modes, which can be selected via its pin-2 PROGRAMME input, as follows.

(1) If pin-2 is left open, the circuit gives 'standard' dimming operation in which a very brief input touch makes the lamp change state (from off to a remembered on state, or vice versa), but a sustained (greater than 400ms) 'dimming' input puts the IC into a ramping mode in which the lamp power slowly ramps up and down (between 3 percent and 97 percent of maximum) until the input is released, at which point the prevailing power level is held and 'remembered'; the ramp direction reverses on alternate 'dimming' touches.

(2) If pin-2 is shorted to pin-7, the lamp goes to maximum brightness when switched on, and in dimming operations the lamp starts at minimum brightness and then slowly ramps up and down until the sensor is released; the ramping direction does not reverse on successive dimming operations.

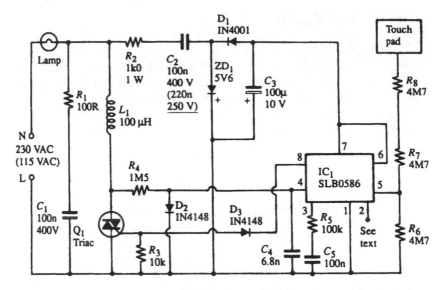

Figure 4.37 *Basic SLB0586 lamp dimmer circuit, with touch-sensitive control*

(3) If pin-2 is shorted to pin-1, the lamp operation is like that just described, except that the ramping direction reverses on successive dimming operations.

Note that the 'touch pads' used with this circuit can be simple strips of conductive material, and that several touch pads can be wired in parallel if desired, enabling the dimmer to be operated from several different points. The circuit can also be activated via any desired number of conventional push-button switches connected in parallel, or via a remote optocoupled input, by using the basic connections shown in *Figure 4.38* or *4.39*. Here, the pin-6 to pin-7 connection is broken and replaced by the R9–R10 divider. The push-button control switch (PB1) is connected between

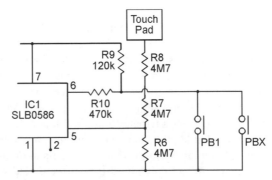

Figure 4.38 *Push-button control applied to the* Figure 4.37 *circuit*

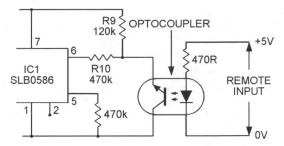

Figure 4.39 *Optocoupler operation of the SLB0586 lamp dimmer circuit*

the R9–R10 junction and pin-1 ('live') of the IC. Any desired number of push-button control switches can be connected in parallel. If the 'touch control' facility is not needed, R7–R8 can be eliminated and R6 can be reduced to 470k, as shown in *Figure 4.39*, which shows how to modify the basic circuit for external control via an optocoupler, without the use of the touch control facility.

Universal-motor controllers

Domestic appliances such as electric drills and sanders, sewing machines and food mixers, etc., are almost invariably powered by series-wound 'universal' electric motors (so called because they can operate from either AC or DC supplies). When operating, these motors produce a back-e.m.f. that is proportional to the motor speed. The effective voltage applied to such motors is equal to the true applied voltage minus the motor's back-e.m.f., which is directly proportional to the motor speed; this fact results in a degree of self-regulation of universal motor speed, since any increase in motor loading tends to reduce the speed and back-e.m.f., thereby increasing the effective applied voltage and causing the motor speed to rise towards its original value.

Most universal motors are designed to give single-speed operation. Triac phase-controlled circuits can easily be used to provide these motors with fully variable speed control, but give rather poor self-regulation under variable loading conditions; a suitable 'diac plus phase-delay' circuit is shown in *Figure 4.40*. This circuit is particularly useful for controlling lightly loaded appliances such as food mixers and sewing machines, for example.

Electric drills and sanders, etc., are subject to very heavy load variations, and are not suitable for control via the *Figure 4.40* circuit. Instead, the variable speed-regulator circuit of *Figure 4.41* should be used. This circuit uses a SCR as the control element and feeds half-wave power to the motor (this results in only a 20 percent or so reduction in maximum available speed/power), but in the OFF half-cycles the back-e.m.f. of the motor is sensed by the SCR and used to give automatic adjustment of the next gating pulse, to give automatic speed regulation. The

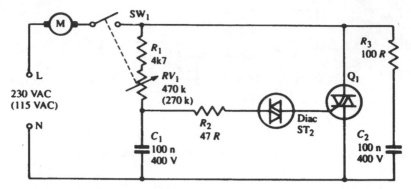

Figure 4.40   *Universal-motor speed controller, for use with lightly loaded appliances (food mixers, sewing machines, etc.)*

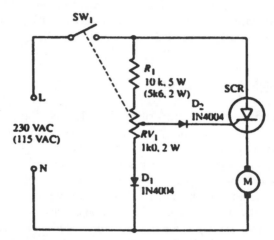

Figure 4.41   *Self-regulating universal-motor speed controller, for use with electric drills and sanders, etc*

R1–RV1–D1 network provides only 90° of phase adjustment, so all motor pulses have minimum durations of 90° and provide very high torque. At low speeds the circuit goes into a 'skip cycling' mode, in which power pulses are provided intermittently, to suit motor loading conditions. The circuit provides particularly high torque under low-speed conditions.

5

DC power control circuits

The DC power feed to loads such as lamps, heaters, buzzers and bells, etc., can easily be controlled via unidirectional solid-state devices such as bipolar transistors, power MOSFETs, or SCRs, which can be used in either the 'static' mode to give a simple ON or OFF switch-control action, or in the pulsed mode to give either a variable power control action (as already described in Chapter 1) or a simple sound-generation action. This chapter shows a variety of ways of using these devices to control such loads (DC motor control circuits are separately described in Chapter 6).

Power-switching basics

Three basic types of solid-state device are available for use in DC power control applications, these being the bipolar transistor, the power MOSFET, and the SCR; each of these devices offers its own particular set of advantages and disadvantages. Consider first the ordinary bipolar transistor.

In most DC power switching applications the bipolar transistor is wired in the common-emitter mode, as shown in *Figures 5.1* and *5.2*. In the case of the npn

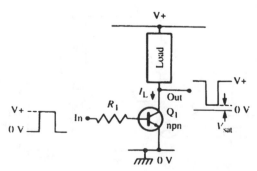

Figure 5.1 *The npn transistor acts as a load-current 'sink'*

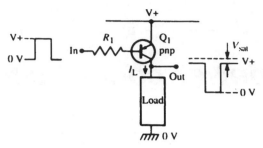

Figure 5.2   *The pnp transistor acts as a load-current 'source'*

device (*Figure 5.1*), the load is wired between Q1 collector and the positive supply rail, and the transistor acts as a current 'sink' (current flows *into* the collector via the load). In the case of the pnp device (*Figure 5.2*), the load is wired between Q1 collector and the negative supply rail, and the transistor acts as a current 'source' (current flows *from* the collector into the load).

The main advantage of the common-emitter configuration is that it offers a very low saturation or 'loss' voltage (typically 200mV to 400mV); the main disadvantage is that it offers fairly low overall current and power gains (typically 100:1). In practice, these gains can easily be increased to 10 000:1 (without increasing the saturation voltage) by either cascading a couple of common-emitter stages, as in *Figure 5.3*, or by wiring a pair of transistors in the super-alpha mode, as shown in *Figure 5.4*.

If really high current/power gains are required, they can be obtained by using a power MOSFET as the DC power switch; these devices have a near-infinite DC input impedance, and thus give near-infinite current and power gains. One of the most popular types of power MOSFET is the VFET family of enhancement mode devices from Siliconix, and *Figure 5.5* shows one of these devices, the VN66AF (which incorporates a built-in input-protection zener diode), used in this mode.

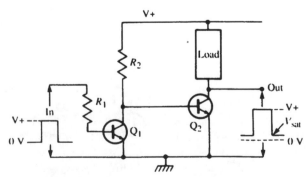

Figure 5.3   *High-gain transistor switch using cascaded npn common-emitter stages*

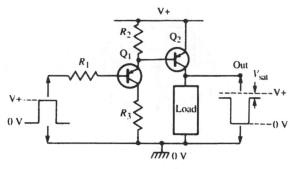

Figure 5.4 *High-gain transistor switch using a super-alpha pnp pair*

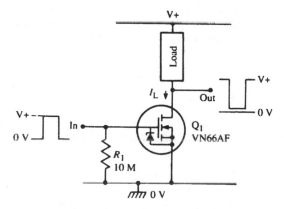

Figure 5.5 *VMOS FETs offer near-infinite current/power gain*

Note that the VN66AF can pass maximum currents of about 2A, and has a typical saturation resistance of about 2R0. Thus, if the device is used with a 12V supply, it will give a saturation voltage of about 80mV when used with a 300R load, or 1.091 volts when used with a 20R load, etc.

The third type of solid-state power switching device is the SCR, which is of particular value in controlling self-interrupting DC loads such as bells, buzzers or sirens; *Figure 5.6* shows the basic circuit. Usually, these loads comprise a solenoid and an activating switch wired in series to give an auto-switching action in which the solenoid first shoots forward via the closed switch, and in doing so forces the switch to open, thus making the solenoid fall back and reclose the switch, thus restarting the action, and so on.

Thus, the basic *Figure 5.6* SCR circuit effectively gives a non-latching load-driving action, since the SCR automatically unlatches each time the load self-interrupts; the load and SCR are thus active only while gate current is applied to the SCR. The circuit can be made fully self-latching, if desired, by shunting the load with resistor R3, as shown, so that the SCR anode current does not fall below the

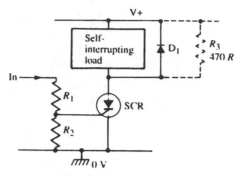

Figure 5.6   *The SCR DC switch offers high-power gain*

SCR's minimum holding value as the load self-interrupts. Note that SCRs offer typical gate-to-anode current gains of about 5000:1, but give saturation voltage values of about 800mV.

Load-type basics

When designing DC switching circuits, some thought must be given to the type of load to be controlled and to its possible harmful effects on the solid-state switching circuity. The most important points to note here are as follows.

When driving filament lamps, note that these devices have a cold or 'switch-on' resistance that is typically one-twelfth of the normal hot or 'running' value, and thus pass switch-on currents that are twelve times greater than the normal running value. Usually, a solid-state switch used for controlling a 500mA lamp must have a surge rating of at least 2A.

When driving highly inductive loads such as relays, solenoids, bells, buzzers, speakers and electric motors, etc., note that these devices can generate very large

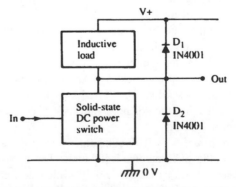

Figure 5.7   *When driving inductive loads, the solid-state switch must be protected via damping diodes*

back-e.m.f.s at the moment of current switch-off, and solid-state power switches need protection against damage from this source. *Figure 5.7* shows how full protection can be provided by using ordinary silicon diodes to 'damp' the back-e.m.f. Here, D1 stops the voltage from swinging more than a few hundred mV above the positive supply rail value, and D2 stops it from swinging significantly below the zero-volts rail. In practice, it is often adequate to provide only partial protection, by using only D1, as in the case of the SCR circuit of *Figure 5.6*.

Finally, when driving loads that are electrically very noisy (such as bells, buzzers, and electric motors, etc.), note that the loads may require damping via small ceramic capacitors, to minimise RFI generation, and that the power supply to the switch-driving circuitry may need extensive ripple decoupling.

LED 'flasher' circuits

Simple 'static' DC ON/OFF lamp or LED power-switching actions are usually best carried out via conventional electromechanical switches or relays, which give minimal voltage loss but generate slight switching RFI. Repetitive DC switching, such as in lamp or LED flasher (pulser) circuits, however, is often best carried out via solid-state power switches, to minimise RFI problems, and *Figures 5.8* to *5.13* show a variety of practical circuits of this type.

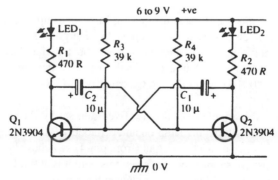

Figure 5.8 *Transistor 2-LED flasher circuit*

Figure 5.8 shows a transistor 2-LED flasher circuit that operates at about 1 flash-per-second and turns one LED on as the other turns off, and vice versa. The circuit is wired as an astable multivibrator, with its timing controlled via C1–R3 and C2–R4, and the LEDs and their current limiters (R1 and R2) are used as the transistor collector loads. This circuit can be converted to single-LED operation by replacing the unwanted LED with a short circuit.

Figure 5.9 shows an IC version of the 2-LED flasher based on the faithful old 555 timer chip or its more modern CMOS counterpart, the 7555. The IC is wired in the

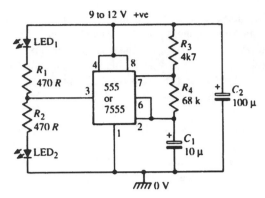

Figure 5.9   *IC 2-LED flasher circuit*

astable mode, with its time constant determined by C1 and R4, and the action is such that IC output pin-3 alternately switches between the ground and positive supply voltage levels, alternately shorting out and disabling one or other of the two LEDs. The circuit can be converted to single-LED operation by omitting the unwanted LED and its associated current-limiting resistor.

Lamp flasher circuits

The above two circuits are suitable for driving LED loads up to only a few tens of mA; greater output currents can be obtained by interposing a suitable transistor 'booster' stage between the astable output and the external load, as shown in *Figures 5.10* to *5.13*. In each of these circuits the astable is designed around one-half of an inexpensive 4001B CMOS IC, and Q2 can drive a 12V lamp load at maximum currents up to 2A.

In *Figure 5.10*, the astable operates at a fixed rate set by R1 and non-polarized (NP) capacitor C1, and turns the lamp on and off about 40 times per minute (i.e. at 1.5 seconds per flash). The astable output (from pin-4) switches alternately between 0V and the full positive supply voltage; when this output is at 0V, Q1–Q2 are driven on via R2, and the lamp is illuminated, but when the output is fully positive Q1–Q2 are cut off, and the lamp is also off. Note that this circuit drives a lamp that has one side tied to the zero-volts rail.

The basic *Figure 5.10* circuit can be usefully modified in a number of ways, as shown in *Figure 5.11*. First, the astable frequency can be made variable by replacing R1 with a series-connected 470k resistor and a variable 1M0 resistor, thus making the flash rate variable between 27 and 80 times per minute. Second, the output can be used to drive lamps that have one side tied to the positive supply rail by using the alternative output stage shown in the diagram. Finally, the astable can, if desired, be turned on and off via SW1 and the high-impedance pin-1 'gate' pin of the IC1a, as shown, rather than via a switch wired in series with the positive supply rail.

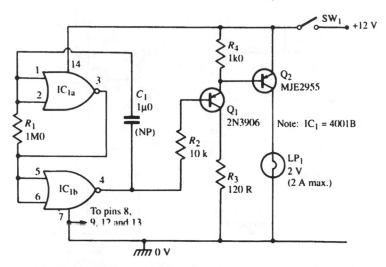

Figure 5.10 *Simple DC lamp flasher*

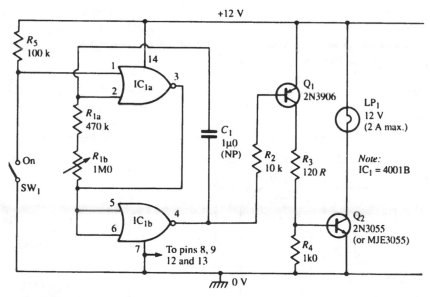

Figure 5.11 *Modified DC lamp flasher circuit*

The above two astable circuits generate 1:1 duty cycles or mark–space ratios, and thus turn the lamp on and off for equal periods. *Figure 5.12* show how the basic circuit can be modified to give a programmed duty cycle (PDC) so that, for example, the lamp gives a 0.75 second flash once every 8.25 seconds, thus giving a 1:10 duty cycle and giving great current economy as an emergency lamp flasher.

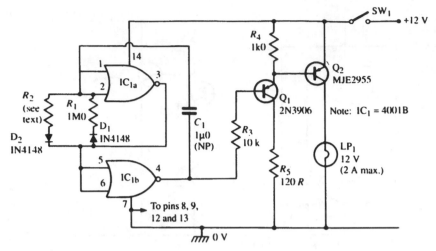

Figure 5.12   *Programmed-duty-cycle (PDC) lamp flasher circuit*

The lamp's ON time is controlled by D1–R1 and is fixed at about 0.75 seconds, but the OFF time is controlled by D2–R2 and can be varied over a wide range; the R2 value can vary from a few kilohms to tens of megohms, and gives an OFF time of 0.75 seconds at 1M0, or 7.5 seconds at 10M, etc.

Figure 5.13 shows how the above circuit can be modified for use with the alternative type of output stage, and with direct astable ON/OFF gating. This

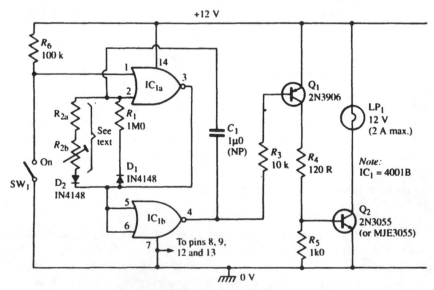

Figure 5.13   *Gated and modified PDC lamp flasher*

diagram also shows how R2 can be replaced by a fixed and a variable resistor in series, so that the OFF time of the lamp is made variable. No specific values are shown for these two resistors, but note that R2a determines the minimum value of the OFF time, and R2b determines the maximum OFF value.

Automatic flashers

Each of the above four lamp flasher ciruits are manually activated; *Figures 5.14* and *5.15* show how the basic and the PCD circuits can be modified so that they activate automatically when the ambient light level falls below a pre-set value.

Both of these circuits operate in the same basic way; in each case, the astable is wired in the 'gated' mode, but has its gate (pin-1) signal applied in the form of a variable voltage that is generated via light-sensitive potential divider LDR RV1. The action of the astable is such that it is disabled, with its pin-4 output driven high, when this gate voltage exceeds a 'threshold' value of approximately half-supply volts, but is automatically enabled when this voltage is below the threshold value.

In these circuits, the light-sensitive LDR generates a low resistance under bright conditions, and a high resistance under dark ones, and the LDR–RV1 divider thus generates a high voltage when it is bright, and a low voltage when it is dark. In practice, RV1 enables this voltage to be adjusted so that it precisely matches the IC's 'threshold' value at the desired light 'trip' level. Consequently, when RV1 is suitably adjusted, these lamp flasher circuits turn on automatically when the light level falls below the desired trip level, and turn off again when it rises above that level.

The LDRs used in these circuits can be any cadmium-sulphide photocells with resistances greater than a few thousand ohms at the required turn-on light levels;

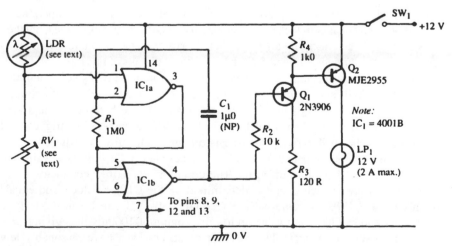

Figure 5.14 *Automatic (dark-activated) DC lamp flasher*

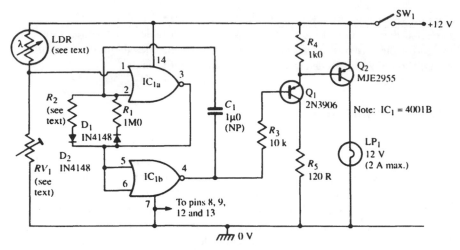

Figure 5.15   *Automatic PDC lamp flasher*

RV1 should have a maximum value roughly double that of the LDR under the turn-on condition. When building these circuits, note that the LDR faces must be shielded from the light of the flasher lamps, so that the LDRs respond to ambient light level but are unaffected by the lamp-flashing action. Also note that master ON/OFF switch SW1 is wired in series with the supply line of each circuit, so that each circuit can be disabled when put away in a dark storage area.

One-shot lamp drivers

Another useful type of DC lamp-driving power control circuit is the one-shot or auto-turn-off design, which turns the lamp on as soon as a START button is pressed, but then turns it off again automatically after a pre-set interval variable from a few seconds to several minutes. *Figures 5.16* and *5.17* show practical circuits of this type.

Both these circuits operate in the same basic way, and have two gates of a 4001B CMOS IC wired as a simple manually triggered monostable multivibrator. Normally, the monostable output is low, so Q1 and Q2 are both cut off and zero power is fed to the lamp. When SW1 is briefly closed the monostable triggers and its output switches high, driving the transistors and lamp fully on. After a pre-set delay, the monostable output automatically switches back to the low state again, and the transistors and lamp turn off again. The circuit action is then complete.

The time delay of each circuit is determined by the C1–R1 values, and equals roughly 0.5sec/µF of C1 value when R1 equals 680k. C1 can be an electrolytic component and can have any value up to a maximum of 1000µF, thus giving time delays up to several minutes; this capacitor must, however, have reasonably low leakage characteristics.

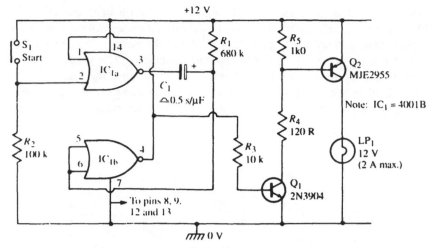

Figure 5.16   *Auto-turn-off time-controlled DC lamp driver*

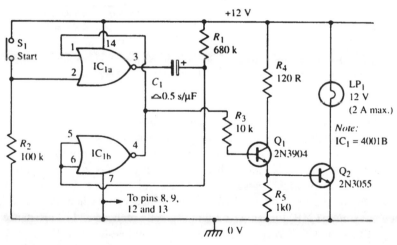

Figure 5.17   *Alternative auto-turn-off lamp driver circuit*

Note that these two circuits can give control of lamps that have one side taken to either the positive or the zero voltage rail, but that the transistor output stages differ from those described earlier.

DC lamp-dimmer circuits

The ordinary filament lamp consists of a coil of tungsten wire (the filament) suspended in a vacuum- or gas-filled glass envelope and externally connected via a pair of metal terminals; the filament runs white hot when energised via a suitable AC or DC voltage, thus generating a bright white light; the filament resistance varies with the filament temperature. *Figure 5.18(a)* shows the standard symbols used to represent such a lamp, and *Figure 5.18(b)* shows the typical filament resistance variation that occurs in a 12V, 12W lamp; the resistance is 12R when the filament is operating at its normal 'white' heat, but is only 1R0 when the filament is cold. This 12:1 resistance variation is typical of all filament lamps, and causes them to have switch-on 'inrush' current values about twelve times greater than the normal 'running' values.

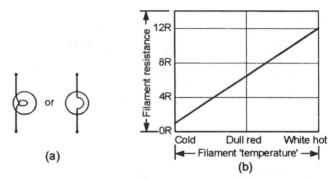

Figure 5.18 *Filament lamp symbols* (a) *and graph* (b) *showing typical variation in filament resistance with filament temperature for a 12V, 12W lamp*

One easy way to vary the brilliance of a DC-powered lamp is to wire a rheostat and a ganged switch in series with the lamp as shown in *Figure 5.19*, in which RV1 has a maximum resistance value double that of the 'hot' resistance value of the 12V, 12W lamp of *Figure 5.18*. Here, when RV1's resistance is zero, the full 12V is applied to the lamp, which presents a resistance of 12R and dissipates 12W (as shown in *Figure 5.20(a)*) and thus burns at full brilliance, but when RV1's resistance is set to maximum (24R) the lamp's current is greatly reduced and its resistance falls to only 3R0, so only 0.59W are dissipated in the lamp (as shown in *Figure 5.20(b)*), which thus produces very little light output. This simple circuit thus enables the

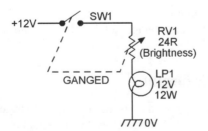

Figure 5.19 *Rheostat brightness control circuit*

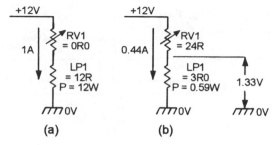

Figure 5.20 *Equivalent of the* Figure 5.19 *circuit at (a) maximum and (b) minimum brilliance levels*

lamp's power dissipation to be varied over a 20:1 range, and enables its brilliance to be varied over an even greater range.

Circuits of the *Figure 5.19* type are often used as low-power lamp dimmers in motor vehicles, but waste a lot of power in RV1, which must have a substantial power rating and be capable of handling the cold currents of the lamp. *Figure 5.21* shows an alternative DC lamp-dimmer circuit, which dissipates negligible power in RV1.

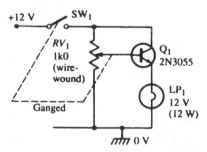

Figure 5.21 *'Variable-voltage' DC lamp brightness control circuit*

In *Figure 5.21*, RV1 acts as a variable potential divider which applies an input voltage to the base of emitter follower Q1, which buffers (power boosts) this voltage and applies it to the lamp. RV1 thus enables the lamp voltage (and thus its brilliance) to be fully varied from zero to maximum. Disadvantages of this simple circuit are that it is very inefficient, since all unwanted power is 'lost' across Q1, and Q1 needs a large power rating and must be capable of handling the cold current of the lamp.

Figure 5.22 illustrates the basic principles of switched-mode variable power control. Here, an electronic switch (SW1) is wired in series with the lamp and can be opened and closed via a pulse-generator waveform. When this pulse is high, SW1 is closed and power is fed to the lamp; when the pulse is low SW1 is open, and power is not fed to the lamp.

The important thing to note about the *Figure 5.22* pulse generator is that it generates a waveform with a fixed frame width but with a variable M/S (on/off)

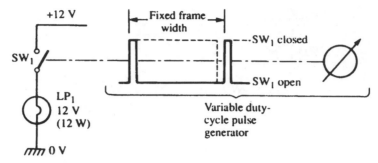

Figure 5.22   *Basic 'switched-mode' DC lamp brightness control circuit*

ratio or duty cycle, thereby enabling the *mean* lamp voltage to be varied. Typically, the M/S ratio is fully variable from 1:20 to 20:1, enabling the mean lamp voltage to be varied from 5 to 95 percent of the DC supply-voltage value.

Because of the inherently long thermal time constant of a filament lamp, its brilliance responds relatively slowly to rapid changes in input power. Consequently, if the frame width of the *Figure 5.22* waveform generator is less than roughly 20ms (i.e. the repetition frequency is greater than 50Hz), the lamp will show no sign of flicker, and its brilliance can be varied by altering the M/S ratio.

Thus, if the M/S ratio of the *Figure 5.22* circuit is set at 20:1, the mean lamp voltage is 11.4V and the consequently hot lamp consumes 10.83W. Alternatively, with the M/S ratio set at 1:20, the mean lamp voltage is only 600mV, so the lamp is virtually cold and consumes a mere 360mW. The lamp power consumption can thus be varied over a 30:1 range via the M/S-ratio control. Note, however, that this wide range of control is obtained with virtually zero power loss within the system, since power is actually controlled by SW1, which is always either fully on or fully off. The switched-mode control system is thus highly efficient (typical efficiency is about 95 percent).

Figure 5.23 shows a practical switched-mode DC lamp dimmer circuit. Here, IC1a and IC1b are wired as an astable multivibrator that operates at a fixed frequency of about 100Hz and has its output fed to the lamp via Q1 and Q2, but has the Mark part of its waveform controlled via C1–D1–R1 and the right-hand part of RV1, and the Space part controlled via C1–D2–R2 and the left-hand part of RV1, thus enabling the M/S ratio to be varied from 1:20 to 20:1 via RV1, which enables the mean lamp power to be varied over a 90:1 range. Note that ON/OFF switch SW1 is ganged to RV1, enabling the circuit to be turned fully off by turning the RV1 brilliance control fully anticlockwise, and that R6-C2 protect IC1 against damage from supply-line transients, thus enabling the circuit to be powered via 12V motor vehicle supplies, etc.

The *Figure 5.23* circuit can be used to control the brilliance of any low-power (up to 24W) 12V filament lamp that has one side wired to the zero volt rail of the power supply. *Figure 5.24* shows how the design can be modified (by using an alternative

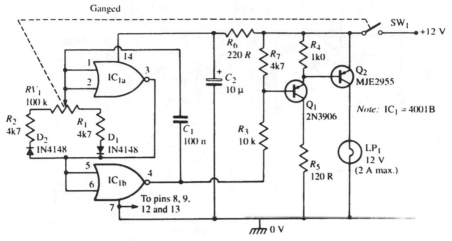

Figure 5.23 *Switched-mode DC lamp dimmer*

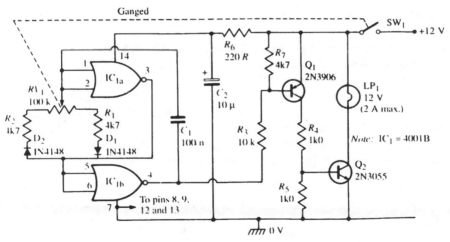

Figure 5.24 *Alternative switched-mode DC lamp dimmer*

output stage) for use with lamps that have one side wired to the positive rail of the power supply.

Finally, *Figure 5.25* shows how the basic switched-mode circuit can be used to efficiently control the brilliance of one (or more) LEDs, at maximum currents up to about 20mA. In this case the two spare gates of IC1 are wired as simple parallel-connected inverters and used to provide a medium-current buffered drive to the LED via current limiter R3. This circuit can be powered from any DC supply in the range 5 to 15V.

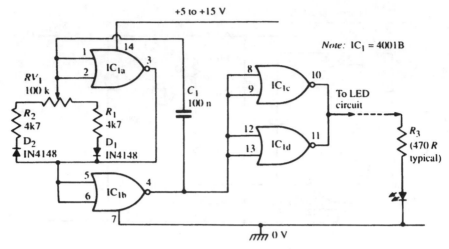

Figure 5.25   *Switched-mode LED brightness-control circuit*

Bell/buzzer alarm circuits

One of the easiest ways of activating a self-interrupting audible alarm device such as a bell or buzzer is via a SCR, and *Figures 5.26* to *5.34* show a selection of circuits of this type. All of these are designed around the inexpensive and readily available type C106 SCR, which can handle mean load currents up to 2.5 amps, needs a gate current of less than 200μA, and has a 'minimum holding current' value of less than 3mA. Note in all cases that the circuit's supply voltage should be about 1.5V greater than the nominal operating voltage of the alarm device used, to compensate for voltage losses across the SCR, and that diode D1 is used to damp the alarm's back-e.m.f.'s.

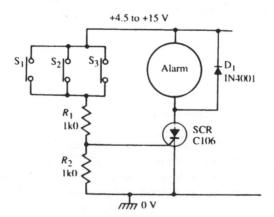

Figure 5.26   *Multi-input non-latching alarm circuit*

Figure 5.26 shows the circuit of a simple non-latching multi-input alarm, in which the alarm activates when any of the S1 to S3 push-button input switches are closed, but stops operating as soon as the switch is released.

Figure 5.27 shows how the above circuit can be converted into a self-latching multi-input 'panic' alarm (which can be activated by the owner if he/she feels immediately threatened while at home) by simply wiring R3 plus normally closed RESET switch S4 in parallel with the alarm device, so that the SCR's main-terminal current does not fall below the 3mA 'minimum holding' value when the alarm self-interrupts. Once this circuit has latched, it can be unlatched again (reset) by briefly opening S4.

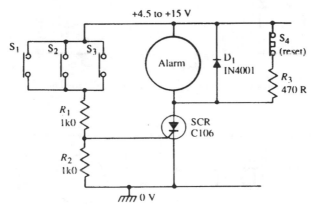

Figure 5.27   *Multi-input self-latching 'panic' alarm*

Note that both of the above circuits pass typical standby currents of only 0.1µA when the alarm is in the OFF condition, and can thus be powered from battery supplies, and that the S1 to S3 switches pass activating currents of only a few mA, and can thus safely be connected to the alarm circuit via considerable lengths of cable.

Figure 5.28 shows how the self-latching circuit can be converted into a simple burglar alarm system, complete with the 'panic' facility. Here, the alarm can be activated by briefly opening any of the series-connected normally closed S1 to S3 'burglar alarm' switches (which can take the form of microswitches that are activated by the action of opening doors or windows, etc.), or by briefly closing any of the parallel-connected normally open 'panic' switches. Note that this circuit passes a typical standby current of 0.5mA (via R1) when powered from a 6V supply, and that C1 acts as a noise-suppressing capacitor that ensures that the alarm will only operate if the S1 to S3 switches are held open for more than a millisecond or so, thus enhancing the circuit's reliability.

The standby current of the burglar alarm circuit can be reduced to a mere 1.4µA (at 6V) by modifying it as shown in *Figure 5.29*, where Q1 and Q2 are connected in the Darlington mode and wired as a common-emitter amplifier, which inverts and boosts the R1-derived 'burglar' signal and then passes it on to the gate of the SCR.

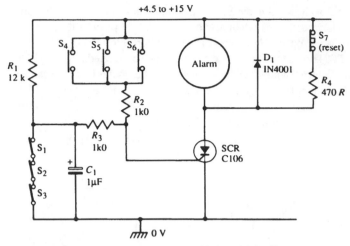

Figure 5.28   *Simple burglar alarm system, with 'panic' facility*

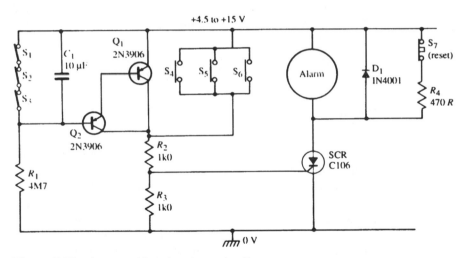

Figure 5.29   *Improved burglar alarm circuit*

Water, light and heat alarms

The basic SCR-driven alarm circuit can be used to indicate the presence of excess water, light, or temperature levels by driving the SCR gate via suitable 'detection' circuitry; *Figures 5.30* to *5.34* show alarm circuits of this type.

The *Figure 5.30* 'water-activated' alarm uses Q1 to effectively increase the SCR's gate sensitivity, and activates when a resistance of less than about 220k appears across the two metal probes. Its operation as a water-activated alarm relies on the

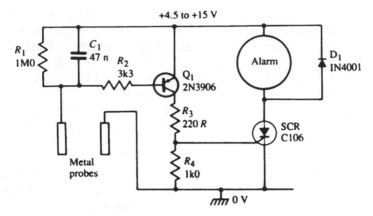

Figure 5.30 *Water-activated alarm*

fact that the impurities in normal water (and many other liquids and vapours) make it act as a conductive medium with a moderately low electrical resistance, which thus causes the alarm to activate when water comes into contact with both probes simultaneously. C1 is used to suppress unwanted ac signal pick-up, and R2 limits Q1's base current to a safe value. By suitably adjusting the placing of the two metal probes, this circuit can be used to sound an alarm when water rises above a pre-set level in a bath, tank, or cistern, etc.

The operation of the *Figure 5.31* 'light-activated' alarm is very simple. The LDR and RV1 are wired as a light-sensitive voltage-generating potential divider that has its output buffered via Q1 and fed to the SCR gate via R1; this output is low under dark conditions (when the LDR resistance is high), but goes high under bright conditions (when the LDR resistance is low) and thus drives the SCR and alarm ON.

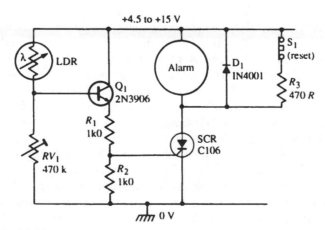

Figure 5.31 *Light-activated alarm*

The precise light-triggering point of the circuit can be pre-set via RV1, and almost any small cadmium sulphide photocell can be used in the LDR position. This circuit can be used to sound an alarm when light enters a normally-dark area such as a drawer or wall safe, etc.

Temperature-activated alarms can be used to indicate either fire or overheat conditions, or frost or underheat conditions. *Figures 5.32* to *5.34* show three such circuits; in each of these designs TH1 can be any n.t.c. thermistor that presents a resistance in the range 1k0 to 20k at the required trigger temperature, and pre-set pot RV1 should have a maximum resistance value roughly double that of TH1 under this trigger condition.

The *Figure 5.32* over-temperature alarm operates as follows. R1–R2 and TH1–RV1 are wired as a Wheatstone bridge in which R1–R2 generates a fixed half-supply 'reference' voltage and TH1–RV1 generates a temperature-sensitive

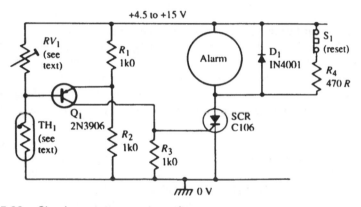

Figure 5.32   *Simple over-temperature alarm*

'variable' voltage, and Q1 is used as a bridge balance detector and SCR gate driver. RV1 is adjusted so that the bridge is normally balanced, with the reference and variable voltages at equal values, at a temperature just below the required trigger value, and under this condition Q1 base and emitter are at equal voltages and Q1 and the SCR are thus cut off. When the TH1 temperature is below this 'balance' value the TH1–RV1 voltage is above the reference value, so Q1 is reverse biased and the SCR remains off, but when the TH1 temperature is significantly above the 'balance' value the TH1–RV1 voltage is below the reference value, so Q1 is forward biased and drives the SCR on, thus sounding the alarm. The precise trigger point of the circuit can be pre-set via RV1.

The action of the above circuit can be reversed, so that the alarm turns on when the temperature falls below a pre-set level, by simply transposing the TH1 and RV1 positions, as shown in the frost or under-temperature alarm circuit of *Figure 5.33*.

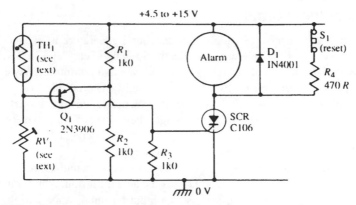

Figure 5.33 *Simple frost or under-temperature alarm*

The above two circuits perform very well, but their precise trigger points are subject to slight variation with changes in Q1 temperature, due to the temperature dependence of the Q1 base-emitter junction characteristics. These circuits are thus not suitable for use in precision applications, unless Q1 and TH1 operate at equal temperatures. This snag can be overcome by using a two-transistor differential detector in place of Q1, as shown in *Figure 5.34*.

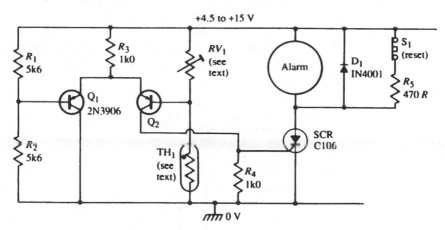

Figure 5.34 *Precision over-temperature alarm*

The *Figure 5.34* circuit is wired as a precision over-temperature alarm. It can be made to function as a precision under-temperature alarm by simply transposing the RV1 and TH1 positions. Note that the circuit is shown without a latching resistor, since sensitive circuits of this type are usually required to sound the alarm only so long as the TH1 temperature is beyond the pre-set limit.

Piezo-electric alarms

Piezo-electric transducers are widely used as sound generators in gadgets such as toys, clocks, watches, calculators, and electronic games, etc., and in a variety of applications where space and operating efficiency are at a premium. They consist of a thin slice of electro-constrictive (piezo) ceramic material plus two electrical contacts, and act as super-efficient electric-to-acoustic power converters when operated in the 1kHz to 5kHz frequency range; they give typical power conversion efficiencies of 50 percent, compared to about 0.5 percent for conventional loudspeakers, and thus act as excellent 'sound makers'.

Piezo-electric transducers are available from a several manufacturers. The Toko PB2720 is fairly typical of the type. It houses the actual transducer in a small 'easy-to-use' plastic-moulded housing, and its two input terminals appear to the outside world as a simple capacitor with a static value of about 20nF and a DC resistance of near infinity. The most effective and cheapest way to drive the device is to feed it with square waves, but in this case the driver must be able to source and sink currents with equal ease and must have a current-limited (short-circuit proof) output. CMOS drivers fit this bill perfectly.

Figures 5.35 and *5.36* show two inexpensive ways of driving the PB2720 (or any similar device) from a 4011B CMOS astable oscillator. Each of these circuits generates a 2kHz monotone signal when in the ON mode, is gated on by a high (logic-1) input, and can use any DC supply in the range 3 to 15V.

In the *Figure 5.35* design, IC1a–IC1b are wired as a 2kHz astable that can be gated on electronically or via push-button switch S1, and IC1c is used as an inverting buffer/amplifier that gives single-ended drive to the PB2720. The signal reaching the PB2720 is thus a squarewave with a peak-to-peak amplitude roughly equal to the supply voltage, and the r.m.s. signal voltage across the load equals roughly 50 percent of the supply line value.

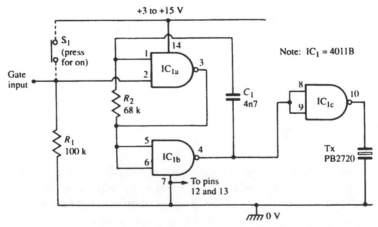

Figure 5.35   *Gated piezo-electric monotone alarm with single-ended output*

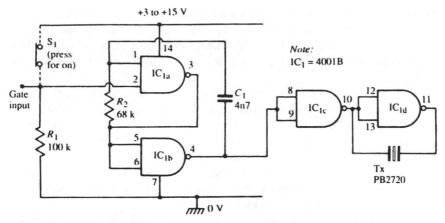

Figure 5.36 *Gated piezo-electric monotone alarm with bridge-drive output*

The *Figure 5.36* design is similar to the above, except that inverting amplifiers IC1c and IC1d are series connected and used to give a 'bridge' drive to the transducer, with anti-phase signals being fed to the two sides of the PB2720. The consequence of this drive technique is that the load (the PB2720) actually sees a squarewave drive voltage with a peak-to-peak value equal to twice the supply

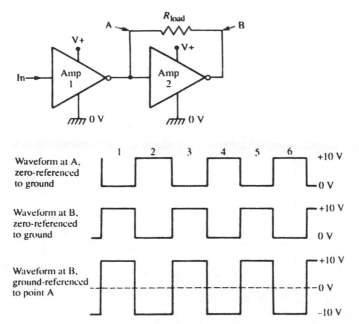

Figure 5.37 *A pair of amplifiers connected in the 'bridge-driving' mode give a power output of $2V^2/R$ watts, i.e. four times the power of a single-ended output*

voltage value, and an r.m.s. voltage equal to the supply value, and thus gives four times more acoustic output power than the *Figure 5.35* design. This action can be understood with the aid of *Figure 5.37*, which shows the waveforms applied to the load from the 'bridge' circuit when it is fed with a 10V peak-to-peak squarewave input signal.

Note in *Figure 5.37* that although waveforms A and B each have peak values of 10V relative to ground, the two signals are in anti-phase (shifted by 180°). Thus, during period 1 of the drive signal, point B is 10V positive to A and is thus seen as being at +10V. In period 2, however, point B is 10V negative to point A, and is seen as being at −10V. Consequently, if point A is regarded as a zero voltage reference point, it can be seen that the point B voltage varies from +10V to −10V between periods 1 and 2, giving a total voltage change of 20V across the load. Similar changes occur in all subsequent waveform periods.

Thus, the load in a 10V bridge-driven circuit sees a total voltage of 20V peak-to-peak, or twice the single-ended input voltage value, as indicated in the diagram. Since doubling the drive voltage results in a doubling of drive current, and power equals the *V–I* product, the bridge-driven circuit thus produces four times more power output than a single-ended circuit.

Alarm circuit variations

Gated CMOS oscillator/driver circuits can be used in a variety of ways to produce useful sounds from the PB2720. A few variations are shown in *Figures 5.38* to *5.40*. *Figure 5.38*, for example, shows how the basic bridge-driving circuit can be modified so that it can be gated on by a low (logic-0) input (rather than by a high one) by simply substituting a 4001B CMOS IC for the 4011B type.

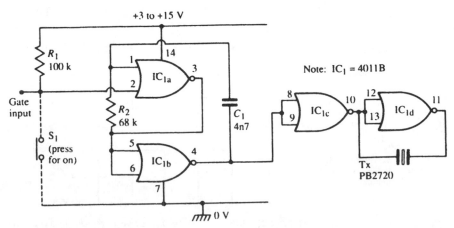

Figure 5.38   *Alternative version of the gated bridge-driving circuit*

Figure 5.39 shows how to use a single 4011B to make a pulsed-tone (bleep-bleep) alarm circuit with direct drive to the PB2720. Here, IC1a–IC1b are wired as a gated 6Hz astable and is used to gate the IC1c–IC1d 2kHz astable on and off. This circuit is gated on by a high input; if low-input gating is wanted, simply swap the 4011B for a 4001B and transpose the positions of S1 and R1.

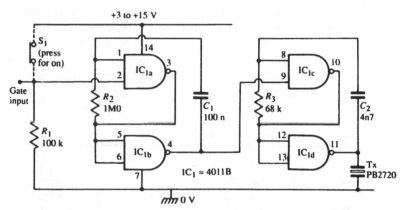

Figure 5.39 *Gated pulsed-tone (6Hz and 2kHz) alarm with piezo output*

Finally, *Figure 5.40* shows a warble-tone (dee-dah-dee-dah) version of the gated alarm which generates a sound similar to a British police car siren and has a bridge-driven output. Here, 1Hz astable IC1a–IC1b is used to modulate the frequency of the IC1c–IC1d astable; the depth of frequency modulation depends on the value of R3, which can have any value from 120k to 1M0.

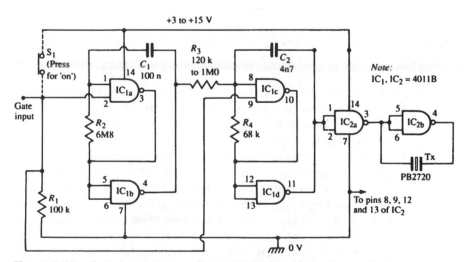

Figure 5.40 *Gated warble-tone alarm with bridge-driven piezo output*

Loudspeaker alarms

The basic CMOS alarm-sound generator astable circuits of *Figures 5.35* to *5.40* can easily be modified to generate acoustic outputs via loudspeakers, thus making greater sound levels available. *Figure 5.41*, for example, shows how the *Figure 5.38* '4001B' gated astable can be operated at 800Hz and used to generate such an output. Here, the astable's pin-4 output is fed to the base of pnp common-emitter 'driver' amplifier Q1 via R3, and Q1 uses the speaker and current-limiting resistor R_x as its collector load. The output of the 4001B astable goes high when gated off, and under this condition the pnp transistor is cut off and consumes zero current; when the astable is active, Q1's output switches on and off at an 800Hz rate and generates an acoustic output via the speaker.

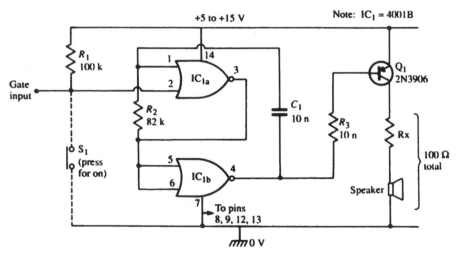

Figure 5.41 *Low-power 800Hz monotone alarm with speaker output*

Note that the 4011B version of the gated CMOS astable (see *Figures 5.35* and *5.36*, etc.) gives a pin-4 output that is grounded when gated off, and this output must thus be fed to the input of an npn driver stage (with a grounded emitter) if a speaker output equal to the above is required.

The basic *Figure 5.41* circuit is intended for low-power applications, and can be used with any speaker in the range 3R0 to 100R and with any supply in the range 5V to 15V. Note that resistor R_x is wired in series with the speaker and must have its value chosen so that the total resistance is roughly 100R, to keep the dissipation of Q1 within acceptable limits. The mean output power level of the circuit depends on the individual values of speaker impedance and supply voltage used, but is usually of the order of only a few tens of milliwatts. Using a 9V supply, for example, the output power to a 15R speaker is about 25mW, and to a 100R speaker is about 160mW.

If desired, the output power of the above circuit can be greatly increased by modifying its output to accept the power booster circuits of *Figures 5.42* or *5.43*. In these circuits, R2 is wired in series with the collector of the existing Q1 alarm output transistor and provides base drive to a one- or two-transistor booster stage, and the alarm's power supply is decoupled from that of the booster via R1–C1. Note that protection diodes are wired across the speakers of these circuits, to prevent the speaker back-e.m.f.s from exceeding the supply rail voltage.

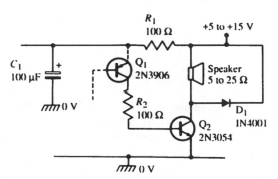

Figure 5.42 *Medium-power (0.25W to 11.25W) booster stage*

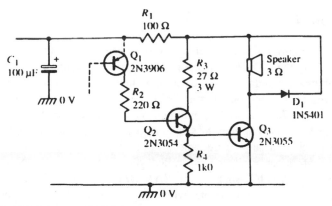

Figure 5.43 *High-power (18W) booster stage*

The *Figure 5.42* booster circuit can be used with any speaker in the range 5R0 to 25R and with any supply from 5V to 15V. The available output power varies from 250mW when a 25R speaker is used with a 5V supply, to 11.25W when a 5R0 speaker is used with a 15V supply. The *Figure 5.43* circuit is designed to operate from a fixed 15V supply and uses a 3R0 speaker, and gives a mean output power of about 18W. Note that, because of transistor leakage currents, these circuits pass typical quiescent current of about 20μA when in the standby mode.

6

DC motor control circuits

One of the most interesting applications of electronic power control techniques is in the control of DC electric motors. These techniques can be used to give precision step rotation or speed/direction control of multi-phase stepper motors, or precision speed regulation or wide-range speed control of permanent magnet DC motors, or precision control of the speed or angular movement of various types of DC servomotor, etc. Practical application circuits of all these types are described in this chapter.

Motor types

There are four major types of DC electric motor that are relevant to this chapter. The first of these is the so-called 'stepper' motor. These motors are provided with a number of phase windings, and each time these are electrically pulsed in the appropriate sequence the output spindle rotates by a precise step angle (typically between 1.8° and 7.5°). Thus, by applying a suitable sequence of pulses, the spindle can be turned a precise number of steps backwards or forwards, or can be made to rotate continuously at any desired speed in either direction.

Stepper motors can easily be controlled via a microprocessor or via dedicated stepper motor driver ICs such as the SAA1027 or SAA1024, for example, and are widely used in all applications where precise amounts of angular movement are required, such as in the movement of robot arms, in daisywheel character selection, or the movement control of the print-head and paper feed in an electronic typewriter, etc.

The most widely used type of DC motor is the permanent magnet commutator type, which is simply designed to rotate at some approximate speed when powered via a particular DC voltage. Motors of this type are often used as fixed-speed drivers in tape/cassette recorders and record/disc players, and as wide-range variable-speed drivers in miniature electric drills and model locomotives. In all of these applications, the motor performance can be greatly enhanced with the aid of electronic control circuitry.

The third type of motor is the so-called 'servomotor', which is an electric motor that is mechanically coupled to a movement-to-data translater such as a shaft-mounted tachogenerator, which gives an output directly proportional to the motor speed, or a potentiometer that is mounted on a gearbox-driven output shaft, thus giving an output that is directly proportional to the shaft position. When one of these motors is coupled into a suitable power control feedback loop, its speed can be precisely locked to that of an external frequency generator, or its shaft movements can be locked to that of an external shaft or control knob.

Servomotors of the tachogenerator type are sometimes used to give precision speed control of high quality record/disc turntables, etc., and servomotors of the 'pot output' type are widely used to give remote-controlled antenna rotation or remote activation of model aircraft control surfaces and engine speed.

The final type of motor is the 2-phase low-voltage 'AC' motor, which is usually driven via a DC-powered low frequency oscillator. Motors of this type are occasionally found in old turntable driving systems.

Stepper motor basics

Stepper motors come in two basic forms, being either variable-reluctance types, or hybrid types. The basic operating principle of the stepper motor can be understood by looking at *Figure 6.1*, which shows four sequential stepping operations of a 4-phase variable-reluctance motor. The stator (body) of this motor has eight inward-projecting teeth, each with a winding or coil that is connected in an opposing sense to the coil on the opposite tooth, to form one phase; there are a total of four phases. When a phase is energised, it generates a magnetic flux that flows from the positive phase tooth to the negative one via the shortest possible magnetic path through the soft-iron rotor; the rotor has (in this case) six projecting teeth, and to minimise this magnetic path the rotor is forced to move so that the nearest pair of its teeth align with those of the energised phase.

Thus, in *Figure 6.1(a)* phase A is energised and the rotor's reference tooth (indicated by a short line) aligns with positive phase tooth A. In *Figure 6.1(b)* phase B is energised, and the rotor is forced to step 15° anticlockwise as the B-phase teeth align with the nearest set of rotor teeth. This process repeats in *Figures 6.1(c)* and *6.1(d)* as phases C and D are sequentially energised, forcing the rotor to step 15° anticlockwise in each case, to give a total of 45° of movement after three phase steps, at which point the rotor's reference tooth has aligned with the phase-D teeth, and the whole A–B–C–D phase stepping sequence can be repeated. Note that the rotor can be moved in an anticlockwise direction by switching the phases in an A–B–C–D sequence, or clockwise by switching them in a D–C–B–A sequence, and that the motor can be spun in either direction by continuously repeating the appropriate set of sequences.

The step length of this type of motor equals $360/p.n$ degrees, where p is the number of phases, and n is the number of rotor teeth. In the case of *Figure 6.1*, this

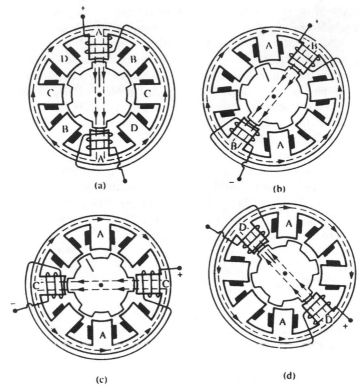

Figure 6.1　*Motor actions in four sequential stepping actions of a 4-phase variable-reluctance stepper motor*

gives a step length of 15°, indicating that a 24-step sequence is needed to complete one motor revolution.

Hybrid stepper motors

The most popular variety of stepper motor is the so-called hybrid type, which gives the same type of stepping action as that of *Figure 6.1*, but differs in its form of construction and operation, in that its rotor incorporates a permanent magnet, and its energising flux flows parallel to the axis of the shaft. Usually, these motors have four phases or coil windings, which may be available via eight independent terminals, as shown in *Figure 6.2(a)*, or via two sets of triple terminals, as shown in *Figure 6.2(b)*. The phases are usually designed for unipolar drive, and must be connected in the correct polarity.

Figure 6.3 shows the basic way of transistor driving a normal 4-phase hybrid stepper motor at its designated voltage rating, and *Figure 6.4* shows the normal full-

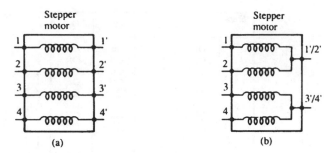

Figure 6.2 *4-phase stepper motors usually have either* (a) *eight or* (b) *six external connection terminals*

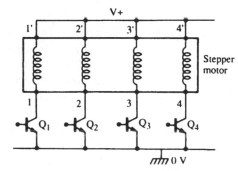

Figure 6.3 *Basic transistor-driven stepper motor circuit*

Step No.	Q_1	Q_2	Q_3	Q_4	
0	On	Off	On	Off	
1	Off	On	On	Off	
2	Off	On	Off	On	
3	On	Off	Off	On	
4	On	Off	On	Off	
5	Off	On	On	Off	

Above sequence repeating

Clockwise ↑ Anticlockwise ↓

Figure 6.4 *Full-step mode sequencing of the* Figure 6.3 *circuit*

step switching sequence. Note that the motor can be repeatedly stepped or rotated clockwise by repeating the 1–2–3–4 sequence or anticlockwise by repeating the 4–3–2–1 sequence. Also note that, in each step, two phases are energised at the same time, but that phases 1 and 2 or 3 and 4 are never both on at the same time.

A useful feature of the 4-phase hybrid motor is that it can also be driven in the 'half-step' mode, in which the rotor advances only a half-step angle at a time, by using a mixture of single and dual phase switching, as shown in *Figure 6.5*.

Step No.	Q_1	Q_2	Q_3	Q_4
0	On	Off	On	Off
1	On	Off	Off	Off
2	On	Off	Off	On
3	Off	Off	Off	On
4	Off	On	Off	On
5	Off	On	Off	Off
6	Off	On	On	Off
7	Off	Off	On	Off
8	On	Off	On	Off
9	On	Off	Off	Off

Clockwise ↑ / Anticlockwise ↑

Above sequence → repeating (rows 8 and 9)

Figure 6.5   *Half-step mode sequencing of the* Figure 6.3 *circuit*

A 4-phase hybrid motor can be operated via a DC supply greater than its designated voltage rating by wiring suitable dropper resistors in series with its phases. Since phases 1 and 2 or 3 and 4 are never both on at the same time, however, each of these pairs of phases can share a single dropper resistor, as shown in *Figure 6.6*. Thus, a 6V, 6R0 motor (1A per phase) can be operated via a 12V supply by giving each resistor a 6R0, 6W rating.

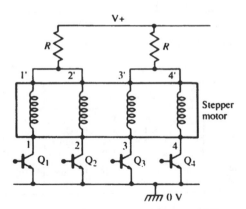

Figure 6.6   *Method of driving a stepper motor from a DC supply greater than its rated value*

The SAA1027 driver IC

A number of dedicated 4-phase stepper motor driver ICs are available, and the best known of these is the SAA1027, which is designed to operate from supplies in the 9.5V to 18V range and to give full-stepping 4-phase motor operation at total output drive currents up to about 500mA.

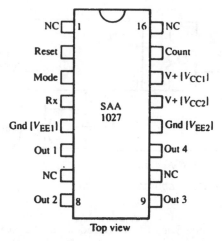

Figure 6.7   *Outline and pin designations of the SAA1027 stepper motor driver IC*

Figure 6.7 shows the outline and pin notations of the SAA1027 IC, *Figure 6.8* shows its internal block diagram, and *Figure 6.9* shows its basic application circuit. Internally, the IC has three buffered inputs which are used to control a synchronous 2-bit (four-state) up/down counter, which has its output fed to a code converter which then uses its four outputs to control (via suitable driver circuitry) four transistor output stages which each operate in the open-collector mode and is protected against damage from motor back-e.m.f.s via an internal collector-to-pin-13 diode.

Note that the IC has two sets of supply rail pins, with one set (pins 13 and 12) feeding the high-current circuitry, and the other (pins 14 and 5) feeding the low

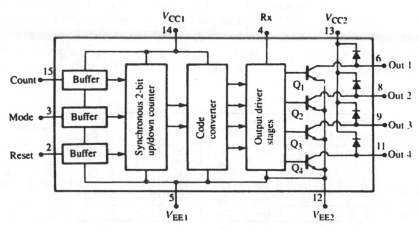

Figure 6.8   *Internal block diagram of the SAA1027*

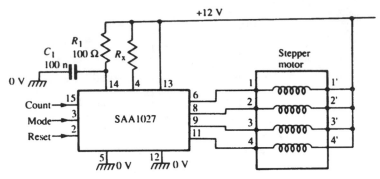

Figure 6.9   *Basic SAA1027 application circuit*

current sections. In use, pins 5 and 12 are grounded, and the positive (usually 12V) rail is fed directly to pin-13 and via decoupling component R1–C1 to pin-14. The positive rail must also be fed to pin-4 via R_x, which determines the maximum drive current capacity of the four output transistors; the appropriate R_x value is given by:

$$R_x = (4E/I) - 60R$$

where E is the supply voltage and I is the desired maximum motor phase current. Thus, when using a 12V supply, R_x needs values of 420R, 180R, or 78R for maximum output currents of 100mA, 200mA, or 350mA respectively.

The SAA1027 IC has three input control terminals, notated COUNT, MODE, and RESET. The RESET terminal is normally biased high, and under this condition the IC's outputs change state each time the COUNT terminal transitions from the low to the high state, as shown in the output sequencing table of *Figure 6.10*. The sequence repeats at 4-step intervals, but can be reset to the zero state at any time by pulling the RESET pin low. The sequence repeats in one direction (normally giving clockwise motor rotation) when the MODE input pin is tied low, and in the other (normally giving anticlockwise motor rotation) when the MODE input pin is tied high.

Figure 6.11 shows a practical drive/test circuit that can be used to activate hybrid 4-phase stepper motors with current rating up to about 300mA. The motor can be

Counting sequence	Mode = low				Mode = High			
	Q_1	Q_2	Q_3	Q_4	Q_1	Q_2	Q_3	Q_4
0	On	Off	On	Off	On	Off	On	Off
1	Off	On	On	Off	On	Off	Off	On
2	Off	On	Off	On	Off	On	Off	On
3	On	Off	Off	On	Off	On	On	Off
0	On	Off	On	Off	On	Off	On	Off
Reset low	On	Off	On	Off	On	Off	On	Off

Above sequence → repeating

Figure 6.10   *SAA1027 output sequencing table*

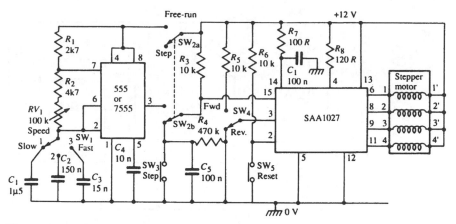

Figure 6.11 *Complete stepper motor drive/test circuit*

manually sequenced one step at a time via SW3 (which is effectively 'debounced' via R4–C5), or automatically via the 555/7555 astable oscillator, by moving SW2 to either the STEP or FREE-RUN position; the motor direction is controlled via SW4, and the stepping sequence can be reset via SW5.

The operating speed of the free-running astable circuit is fully variable via RV1, and is variable in three switch-selected decade ranges via SW1. In the SLOW (1) range, the astable frequency is variable from below 5Hz to about 68Hz via RV1; on a 48-step (7.5° step angle) motor this corresponds to a speed range of 6 to 85 RPM. SW1 ranges 2 and 3 give frequency ranges that are ten and one-hundred times greater than this respectively, and the circuit thus gives a total speed control range of 6 to 8500RPM on a 48-step motor.

Circuit variations

The basic *Figure 6.11* circuit can be varied in a number of ways. *Figure 6.12*, for example, shows how it can be driven via a microprocessor or computer output port with terminal voltages that are below 1V in the logic-0 state and above 3.5V in the logic-1 state. Note that this circuit reverses the normal polarity of the input control signals; thus, the STEP input is pulsed by a high-to-low transition, the stepping sequence is RESET by a high input, and a low MODE input gives forward motor rotation and a high input gives reverse rotation.

The *Figure 6.11* and *6.12* circuits are designed to give maximum output drive currents up to about 300mA. If desired, these outputs can be boosted to a maximum of about 5A by using the circuits of *Figure 6.13* or *6.14*, which each show the additional circuitry needed to drive one of the four output phases of the stepper motor; four such driver stages are needed per motor. The *Figure 6.13* circuit can be used to drive motors with fully independent phase windings, and the *Figure 6.14*

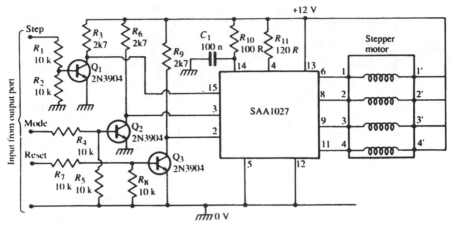

Figure 6.12   *stepper-motor-to-microprocessor interface*

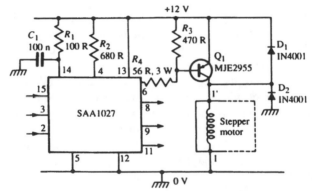

Figure 6.13   *Method of boosting the drive current to stepper motors with independent phase windings*

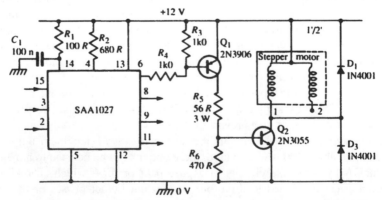

Figure 6.14   *Method of boosting the drive current to stepper motors with coupled phase windings*

design can be used in cases where two windings share a common supply terminal. In both cases, D1 and D2 are used to damp the motor back-e.m.f.s.

Magnet/commutator motor basics

The most widely used type of DC motor is the permanent magnet commutator type, which has a commutator that is designed to rotate when the motor is powered from an appropriate DC voltage. *Figures 6.15(a)* and *6.15(b)* show the circuit symbols and the simplified equivalent circuit of this type of motor.

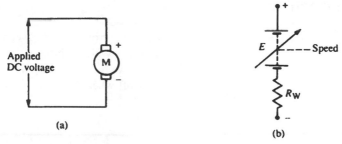

(a)

(b)

Figure 6.15 *Symbol* (a) *and equivalent circuit* (b) *of a permanent-magnet type of DC motor*

The basic action of this motor is such that an applied DC voltage causes a current to flow through sets of armature windings (via commutator segments and pick-up brushes) and generate electromagnetic fields that react with the fields of fixed stator magnets in such a way that the armature is forced to rotate; as it rotates, its interacting fields force it to generate a back-e.m.f. that opposes the applied DC voltage and is directly proportional to the armature speed, thus giving the equivalent circuit of *Figure 6.15(b)*, in which R_w represents the total resistance of the armature windings, etc., and E represents the speed-dependent back-e.m.f. Important points to note about this kind of motor are as follows:

(1) When the motor is loaded by a fixed amount, its speed is directly proportional to supply voltage.
(2) When the motor is powered from a fixed DC supply, its running current is directly proportional to the amount of armature loading.
(3) The motor's effective applied voltage equals the applied DC voltage minus the speed-dependent back-e.m.f. Consequently, when it is powered from a fixed voltage, motor speed tends to self-regulate, since any increase in loading tends to slow the armature, thus reducing the back-e.m.f. and increasing the effective applied voltage, and so on.

(4) The motor current is greatest when the armature is stalled and the back-e.m.f is zero, and then equals V/R_w (where V is the supply voltage); this state naturally occurs under 'start' conditions.

(5) The direction of armature rotation can be reversed by reversing the motor's supply connections.

The main applications of electronic power control to DC motors of this type are in on/off switching control, in direction control, in improved speed self-regulation, and in variable speed control; all of these subjects are dealt with in the next few pages.

On/off switching

A DC motor can be turned on and off by wiring a control switch between the motor and its power supply. This switch can be an ordinary electromechanical type (or a pair of relay contacts), as in *Figure 6.16(a)*, or it can be a switching transistor, as in the *Figure 6.16(b)* circuit, in which the motor is off when the input is low and is on

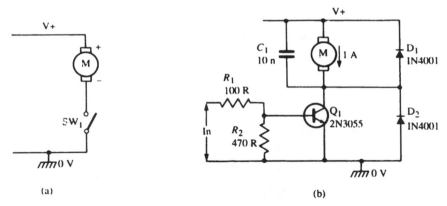

(a)   (b)

Figure 6.16   *ON/OFF motor control using* (a) *electro-mechanical and* (b) *transistor switching*

when the input is high. Note here that diodes D1 and D1 are used to damp the motor's back-e.m.f. that C1 limits unwanted RFI, and that R1 limits Q1's base current to about 52mA with a 6V input, and that under this condition Q1 provides a maximum motor current of 1A.

In the above circuit, Q1's 52mA base current is provided via the external drive circuitry; if desired, the drive current can be reduced to a mere 2mA or so by adding a buffer transistor as shown in *Figure 6.17*; in this case, R3 limits Q1's base current to a safe value.

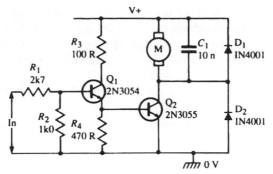

Figure 6.17 *Transistor motor switch with increased sensitivity*

Direction control, using dual supplies

The rotational direction of a permanent magnet DC motor can be reversed by simply reversing the polarity of its supply connections. If the motor is powered via dual (split) supplies, this can be achieved via a single-pole switch connected as in *Figure 6.18*, or via transistor-aided switching by using the circuit of *Figure 6.19*.

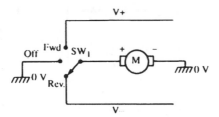

Figure 6.18 *Switched motor-direction control, using dual (split) power supplies*

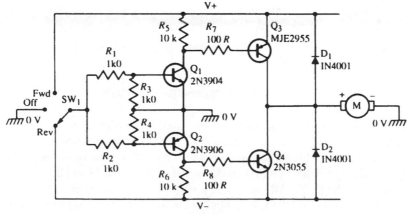

Figure 6.19 *Transistor-switched direction control, using dual supplies*

In *Figure 6.19*, Q1 and Q3 are biased on and Q2 and Q4 are cut off (with Q2's base-emitter junction reverse biased) when SW1 is set to the FORWARD position, and Q2 and Q4 are biased on and Q1 and Q3 are cut off (with Q1's base-emitter junction reverse biased) when SW1 is set to the REVERSE position. Note that if this circuit is used with supply values greater than 12V, diodes must be wired in series with the Q1 and Q2 base-emitter junctions, to protect them against breakdown when reverse biased.

The *Figure 6.19* circuit uses double-ended input switching, and this makes it difficult to replace SW_1 with electronic control circuitry in 'interfacing' applications. *Figure 6.20* shows the design modified to give single-ended input switching control,

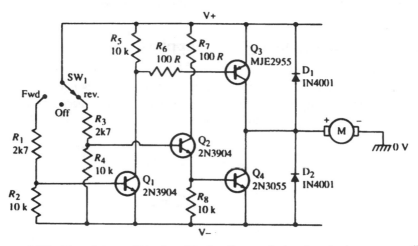

Figure 6.20   *Transistor-switched motor-direction control, using dual power supplies but a single-ended input*

making it easy to replace SW1 with electronic switching. In this circuit, Q1 and Q3 are biased on and Q2 and Q4 are cut off when SW1 is set to the FORWARD position, and Q2 and Q4 are on and Q1 and Q3 are off when SW1 is set to REVERSE.

Direction control, using single-ended supplies

If a permanent magnet DC motor is powered from single-ended supplies, its direction can be controlled via a double-pole switch connected as in *Figure 6.21*, or via a bridge-wired set of switching transistors connected in the basic form shown in *Figure 6.22*. In the latter case, Q1 and Q4 are turned on and Q2 and Q3 are off when SW1 is set to the FORWARD position, and Q2 and Q3 are on and Q1 and Q4 are off when SW1 is set to REVERSE. Diodes D1 to D4 are used to protect the circuit against possible damage from motor back-e.m.f.s, etc.

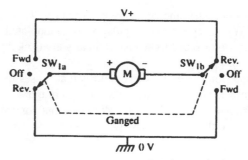

Figure 6.21 *Switched motor-direction control, using a single-ended power supply*

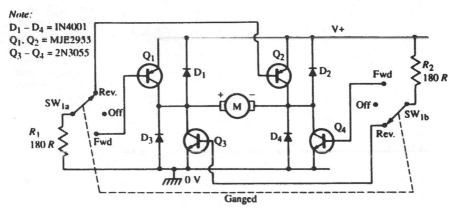

Figure 6.22 *Transistor-switched motor-direction control circuit, using single-ended supplies*

Figure 6.23 shows how the above circuit can be modified to give alternative switching control via independent FORWARD/REVERSE (SW1) and ON/OFF (SW2) switches. A very important point to note about this configuration is that it causes Q1 or Q2 to be turned on at all times, with the ON/OFF action being applied via Q3 or Q4, thus enabling the motor currents to collapse very rapidly (via the

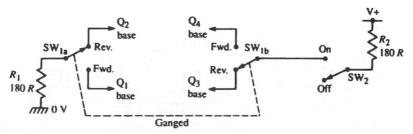

Figure 6.23 *Alternative switching for the* Figure 6.22 *circuit, using separate FWD/ REV (SW1) and ON/OFF (SW2) switches*

Q1–D2 or Q2–D1 loop) when the circuit is switched off. This so-called 'flywheel' action is vital if SW2 is replaced by a pulse-width modulated (PWM) electronic switch, enabling the motor speed to be electronically controlled (this technique will be described shortly).

A weakness of the simple *Figure 6.22* circuit is that it uses fairly high base drive currents, which must be supplied via the switching circuitry. *Figure 6.24* shows a more sensitive version of the circuit, which requires input control currents (to the A, B, C, D terminals) of only a few milliamps.

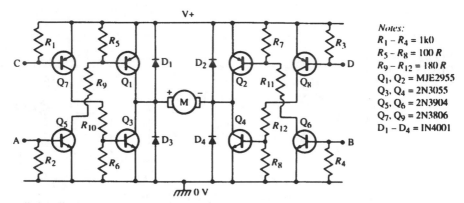

Figure 6.24   *Transistor-switched motor-direction control circuit with increased sensitivity*

The above circuit can be controlled manually via a pair of switches by using the connections shown in *Figure 6.25*, in which SW1 controls the FORWARD/ REVERSE action and SW2 controls the ON/OFF action, or it can be controlled electronically by using the circuit of *Figure 6.26*, in which a 4052B CMOS IC is used as a ganged 2-pole 4-way bilateral switch (see *Figure 3.7*) that can be controlled via logic-0 or logic-1 signals applied to its A or B input pins, to give independent FORWARD/REVERSE and ON/OFF (or PWM speed control) actions.

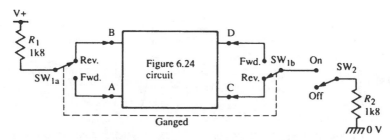

Figure 6.25   *Manual switching connections for use with* Figure 6.24

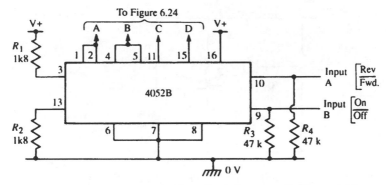

Figure 6.26 *Electronic control switching circuit for use with* Figure 6.24

Figure 6.26 circuit states						Figure 6.24 transistor states			
Inputs		Outputs				Q_1	Q_2	Q_3	Q_4
A|R/F|	B |on/off|	A	B	C	D				
0	0	1	X	X	X	On	X	X	X
0	1	1	X	X	O	On	X	X	On
1	0	X	1	X	X	X	On	X	X
1	1	X	1	0	X	X	On	On	X

X = Off or open circuit

Figure 6.27 *Truth table of the* Figure 6.24 *and* 6.26 *circuits when they are interconnected*

Note that both of these circuits are configured to give the 'flywheel' type of switching action already described. *Figure 6.27* illustrates this point by showing the truth table that occurs when the *Figure 6.24* and *6.26* circuits are interconnected.

Motor speed control

The rotational speed of a DC motor is directly proportional to the mean value of its supply voltage; motor speed can thus be varied by altering either the value of its DC supply voltage, or, if the motor is operated in the switched-supply mode, by varying the mark–space ratio of its supply.

Figure 6.28 shows how variable-voltage speed control can be obtained via variable pot RV1 and compound emitter follower Q1–Q2, which enable the motor's DC voltage to be varied from zero to 12V. This type of circuit gives fairly good speed control and self-regulation at medium to high speeds, but gives very poor low-speed control and slow-start operation, and is thus used mainly in limited-range speed-control applications.

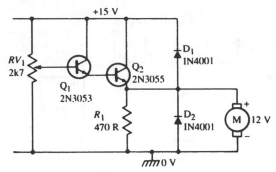

Figure 6.28   *Variable-voltage speed-control of a 12V DC motor*

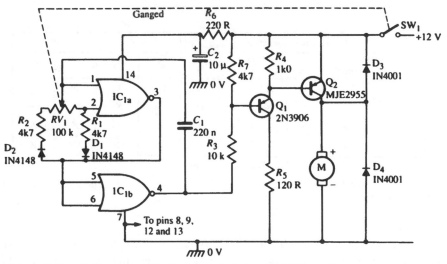

Figure 6.29   *Switched-mode speed-control of a 12V DC motor*

Figure 6.29 shows an example of a switched-mode speed-control circuit. Here, IC1 acts as a 50Hz astable multivibrator that generates a rectangular output with a mark−space ratio fully variable from 20:1 to 1:20 via RV1, and this waveform is fed to the motor via Q1 and Q2. The motor's mean supply voltage (integrated over a 50Hz period) is thus fully variable via RV1, but is applied in the form of high-energy pulses with peak values of 12V; this type of circuit thus gives excellent full-range speed control and generates high torque even at very low speeds; its degree of speed self-regulation is proportional to the mean value of applied voltage.

Model-train speed-controller

Figure 6.30 shows how the switched-mode principle can be used to make an excellent 12V model-train speed-controller that enables speed to be varied smoothly from zero to maximum without jerkiness. The maximum available output current is 1.5A, but the unit incorporates short-circuit sensing and protection circuitry that automatically limits the output current to a mean value of only 100mA if a short occurs on the track. The circuit operates as follows.

The circuit's power line voltage is stepped down via T1 and full-wave (bridge) rectified via BR1, to produce a raw (unsmoothed) DC supply that is fed to the model train (via the track rails) via the series-connected SCR and direction control switch SW3. At the start of each raw DC half-cycle the SCR is off, so DC voltage is applied (via R4 and ZD1) to UJT Q1 and its associated C1–RV1 (etc.) timing circuitry, and C1 starts to charge up until eventually the UJT fires and triggers the SCR; as the

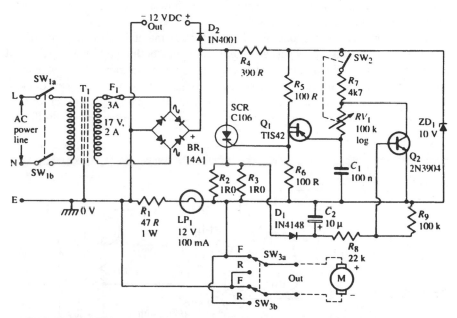

Figure 6.30 *Model train speed-controller circuit with automatic short-circuit protection*

SCR turns on it saturates, removing power from Q1 (which thus resets) and feeding the rest of the power half-cycle to the model train via R2/R3 and SW3. This timing/switching process repeats in each raw DC half-cycle (i.e. at twice the power line frequency), giving a classic phase-triggered power control action that enables the train speed to be varied over a wide range via RV1.

Note that the circuit's output current passes through R2/R3, which generate a proportional output voltage that is peak-detected and stored via D1–C2 and fed to Q2 base via R8–R9. The overall action is such that, because of C2's voltage storing action, Q2 turns on and disables the UJT's timing network (thus preventing the SCR from firing) for several half-cycles if the peak output current exceeds 1.5A. Thus, if a short occurs across the track the half-cycle output current is limited to a peak value of a few amps by the circuit's internal resistance, but the protection circuitry then ensures that the SCR fires only once in (say) every fifteen half-cycles, thus limiting the *mean* output current to roughly 100mA; if the short is still present when the SCR fires again, the protection circuitry is again enabled, but if the short is removed the circuit reverts to its normal operating mode.

An automatic track cleaner

Note that the raw DC output of the above circuit is available (via isolating diode D2) via a pair of output terminals, and can be used to power model railway accessories such as automatic track cleaners, etc.

A major problem with model railways is that of maintaining electrical contact between the train pick-up wheels and the track, which both tend to pick up dirt and oxidisation. This problem can be overcome by feeding the train control signals to the track via a load-sensing high-frequency low-power high-voltage generator or 'track cleaner', which harmlessly cuts its way through any existing dirt or oxidisation. *Figure 6.31* shows an example of such a circuit, and its controller-to-track circuit connections.

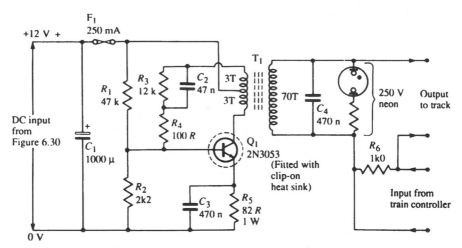

Figure 6.31   *Model railway track cleaner*

The track cleaner is a modified blocking oscillator, tuned to operate at about 100kHz by the inductance of step-up transformer T1 (which is wound on a small ferrite core) and by the values of C2 and C4 (which minimise the unwanted effects of track capacitance). The oscillator generates several hundred volts peak-to-peak on T1 secondary, but at a fairly high impedance (and thus harmless) level; oscillation ceases if the output is heavily loaded.

T1's secondary is wound with fairly heavy gauge wire, through which the train control signals are fed to the track. Thus, when electrical contact is made between a train motor and the track the resulting low impedance kills the oscillator, and only the train control signals reach the track; but if the contact is interrupted by dirt, etc., the resulting high impedance automatically enables the oscillator to work, and the resulting high-frequency high-voltage (plus train control) signals rapidly break through the interruption and re-establish electrical contact.

Note that a neon lamp (plus resistor) is wired across T1 secondary, and illuminates when the track cleaner is active, thus indicating loss of track contact. R6 ensures that only a very small part of the oscillator output voltage can be fed to the train controller terminals when the cleaner is active.

Motor speed regulation

Motor speed regulators are meant to keep motor speed fairly constant in spite of wide variations in the control circuit's supply voltage and in the motor loading conditions, etc. *Figure 6.32* shows an example of a regulator circuit that is simply designed to keep the motor's *applied voltage* constant in spite of wide variations in supply voltage and temperature. It is designed around a 317K 3-terminal variable

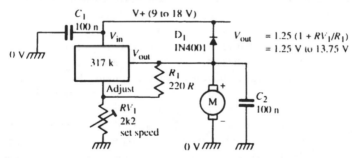

Figure 6.32 *Simple motor-speed controller/regulator*

voltage regulator IC which (when fitted to a suitable heat sink) can supply output currents up to 1.5A and has an output that is fully protected against short-circuit and overload conditions. With the component values shown, the output is fully variable from 1.25V to 13.75V via RV1, provided that the supply voltage is at least 3 volts greater than the desired output value.

Figure 6.33 shows a popular type of regulator circuit that is widely used (with minor variations) in cassette recorders, and compensates for variations in both battery voltage and motor loading conditions. The motor current is controlled via series transistor Q1 and is monitored via R2, and Q1 is controlled via Q2. The diagram shows the circuit voltages obtained when the motor is operating at 6V and is drawing 100mA. Note that Q2's emitter is biased 1.2V below the motor value via

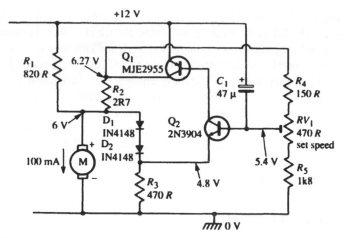

Figure 6.33　*Popular motor-speed regulator circuit*

D1–D2–R3, and Q1's base is set at a fraction of the Q1-collector value via R4–RV1–R5; all changes in motor or Q1 collector voltages thus affect Q2's emitter and base bias values, but the changes are always least on the base.

Thus, any drop in the circuit's supply voltage tends to decreases the motor's voltage and make Q2's emitter fall further below its base value, thus driving Q2 and Q1 harder on and self-compensating for the supply reduction. Similarly, any increase in motor loading tends to slow the motor and increase the motor current and the R2 volt drop, thus raising the relative value of Q1's collector voltage and making Q2's base voltage rise further above that of the emitter, thus driving Q2 and Q1 harder on and increasing the motor drive, to self-compensate for the increased motor loading. This simple circuit therefore gives automatic regulation of motor speed to compensate for variations in supply voltage or loading conditions; some thermal compensation is also given via D1 and D2. The motor speed can be varied over a limited range via RV1.

Finally, *Figure 6.34* shows a high-performance regulator circuit that can be used in wide-range variable-speed applications, such as controlling 12V DC mini-drills, etc. Here, the motor is again powered via the output of a 317K 3-terminal variable voltage regulator IC, but in this case the motor current is monitored via R5–RV2, which feed a proportional voltage to the input of the IC2–Q1 non-inverting DC amplifier, to generate a Q1 emitter voltage directly proportional to the motor's load current.

Now, the output voltage of this circuit equals the normal output value of the 317K IC (which is variable from 1.25V to 13.75V via RV1) plus the voltage on Q1's emitter; consequently, any increase in motor loading makes the circuit's output voltage rise, to automatically increase the motor drive and hold its speed reasonably constant. To initially set up the circuit, simply set the motor speed to about one-third of maximum via RV1, then lightly load the motor and set RV2 so that the speed remains similar in both loaded and unloaded states.

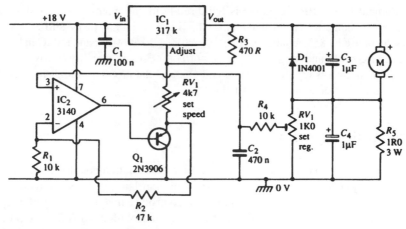

Figure 6.34 *High-performance variable-speed regulator circuit*

Two-phase motor driver

Two-phase (AC) motors are synchronous machines, and low-voltage versions are sometimes used as precision phonograph turntable drivers. *Figure 6.35* shows a circuit that can drive 8R0 motor windings at up to 3W each, at frequencies between 45Hz and 65Hz. The circuit is designed around an LM377 dual 3-watt audio power amplifier IC, driven from a split supply, and operates as follows.

The IC's left-hand half is wired as a Wien bridge oscillator, with frequency variable between 45Hz and 65Hz via RV1 and with amplitude stabilised via RV2

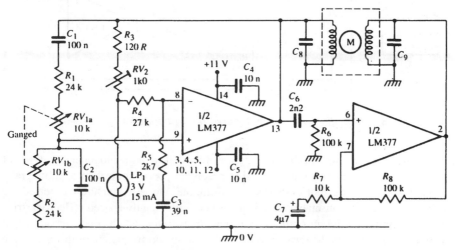

Figure 6.35 *Two-phase motor driver*

and filament lamp LP1; the output is fed directly to one motor phase winding, and to the other via the IC's right-hand half, which acts as an 85° phase shifter (C6–R6 shifts the phase by 85° but attenuates by a factor of 10 at 60Hz, and the IC half gives unity phase shift and a gain of 10 at 60Hz). Circuit stability is assured via decoupling networks C3–R4–R5, C4, and C5, and the motor windings are tuned to a mid-frequency value via C8 and C9.

Servomotor systems

A servomotor is a conventional electric motor with its output coupled (usually via a speed-reduction gearbox) to a movement-to-data translator, such as a potentiometer or a tachogenerator. *Figure 6.36* shows a controller that can be used to give proportional movement (set via RV2) of a servomotor with a pot (RV1) output; the motor can be any 12V to 24V type that draws less than 700mA of current. Here, RV1 and RV2 are wired as a Wheatstone bridge, and the IC (a dual 4W power amplifier) is wired as a bridge-configured motor-driving difference amplifier. The circuit action is such that any movement of RV2 upsets the bridge balance and generates a RV1–RV2 difference voltage that is amplified and fed to the motor, making its shaft rotate and move RV1 to restore the bridge balance; RV1 thus 'tracks' the movement of RV2, which can thus be used to remote-control the shaft position.

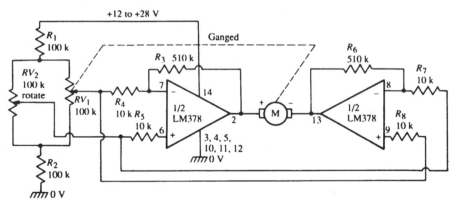

Figure 6.36   *Proportional-movement (servomotor) controller*

Figure 6.37 shows, in block diagram form, how a tachogenerator type of servomotor system can be used to give precision speed control of a phonograph turntable. Here, the motor drives the turntable via a conventional belt drive mechanism, and the turntable edge is patterned with equally spaced reflective strips that are monitored by an optoelectronic tachogenerator which produces an output signal proportional to the turntable speed; the phase and frequency of this signal are compared with that of a precision oscillator, to give an output that is used to control

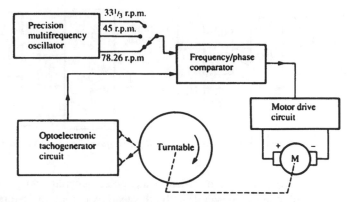

Figure 6.37 *Block diagram showing a tachogenerator servomotor used to give precision speed control of a phonograph turntable*

the motor drive circuit, thus holding the turntable at the precise speed selected by the operator. Some manufacturers produce dedicated ICs for use in this type of application.

One of the best-known types of servomotor is that used in digital proportional remote control systems; these devices actually consist of a special IC plus a motor and a reduction gearbox that drives a pot and gives a mechanical output; *Figure 6.38* shows the block diagram of one of these systems, which is driven via a variable-width (1ms to 2ms) input pulse that is repeated once every 15ms or so (the frame time). The input pulse width controls the position of the servo's mechanical output; at 1ms the servo output may (for example) be full left, at 1.5ms neutral, and at 2ms full right. The system operates as follows.

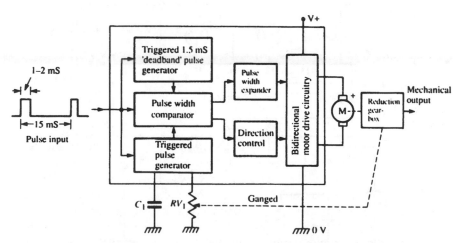

Figure 6.38 *Block diagram showing basic digital proportional-control servomotor system*

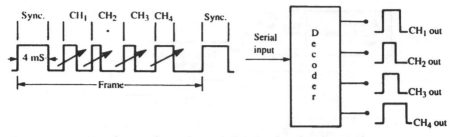

Figure 6.39   *Waveforms of a 4-channel digital proportional-control system*

Each input pulse triggers a 1.5ms 'deadband' pulse generator and a variable-width pulse generator controlled (via RV1) by the gearbox output; these three pulses are fed to a width comparator that gives one output that gives direction control of the motor drive circuitry, and another that (when fed through a pulse-width expander) controls the motor speed, thus making the servomotor's RV1-driving mechanical output rapidly follow any variations in the width of the input pulse.

Servomotors of the above type are usually used in multi-channel remote control systems, as shown in the basic 4-channel system of *Figure 6.39*. Here, a serial data input is fed (via some form of data link) to the input of a suitable decoder; each input frame comprises a 4ms synchronisation pulse followed by four variable-width (1ms

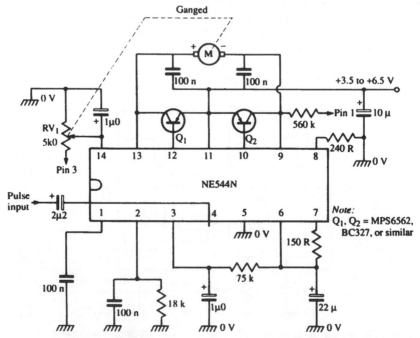

Figure 6.40   *Digital proportional servo driver using the ZN409CE IC*

to 2ms) sequential 'channel' pulses. The decoder simply converts the four channel pulses to parallel form, enabling each pulse to be used to control a servomotor.

Digital servomotor circuits

Digital proportional servomotor units are widely available in both kit and ready-built forms, and are usually designed around either the Ferranti ZN409CE or the Signetics NE544N servo amplifier ICs. *Figures 6.40* and *6.41* show practical application circuits for both of these IC types, with component values suitable for input pulse lengths in the 1ms to 2ms range and frame length of about 18ms nominal.

Finally, to complete this chapter, *Figure 6.42* shows a general-purpose tester for use with the above types of servo. This unit is powered from the servo's supply battery

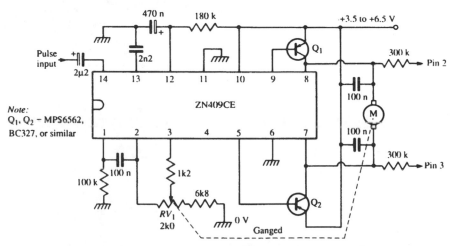

Figure 6.41 *Digital proportional servo driver using the NE544N IC*

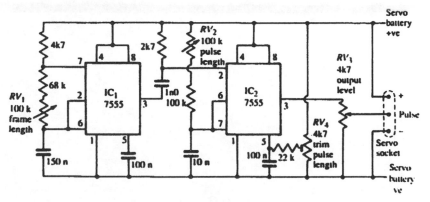

Figure 6.42 *Digital proportional servo tester circuit*

(nominally about 5V) and simply feeds normal input pulses to the servo via a standard servo socket. The frame length is variable from 13ms to 28ms via RV1, and the pulse length is variable from 1ms to 2ms via RV2 and can be trimmed to give a precise 1.5ms mid-scale value via RV4; the output pulse level is variable via RV3.

The circuit is designed around two 7555 ICs, which are CMOS versions of the 555 timer chip and give stable operation at supply values down to 3V. IC1 is configured as a free-running astable multivibrator, and generates the frame times, and its output triggers IC2, which is configured as a monostable multivibrator and generates the output test pulses.

7

Audio power control circuits

Audio power control circuits can, as far as this chapter is concerned, be defined as those that convert an audio input voltage signal into an accurately reproduced acoustic power output (via a loadspeaker), generating a minimum of distortion in the process. A great many circuits of this type, together with audio preamplifier and filter circuits, etc., are already described in two of my other 'Newnes' books, *Audio IC Circuits Manual* and *Audio IC Users Handbook*, which each presents hundreds of audio circuits and diagrams. Consequently, the present chapter simply looks at a small selection of practical IC-based audio power control circuits.

Low-power circuits

Simple audio amplifiers with output powers up to a few hundred milliwatts can be easily and cheaply built using little more than a standard op-amp and a couple of general-purpose transistors. The popular 741 op-amp, for example, can supply peak output currents of at least 10mA and peak output voltage swings of at least 10V onto a 1k0 load when powered from a dual 15V supply, thus giving peak outputs of 100mW under these conditions.

Figure 7.1 shows how the 741 op-amp can be used as a low-level audio power amplifier, using a dual power supply. The external load is direct-coupled between the op-amp output and ground, and the two input terminals are ground referenced. The op-amp is used in the non-inverting mode, and has a voltage gain of ×10 (=R1/R2) and input impedance of 47k (=R3).

Figure 7.2 shows how to use the 741 with a single-ended power supply. Note in this case that the load is ac coupled between the output and ground, and that the output is biased to a quiescent half-supply voltage value (to give maximum output voltage swing) via the R1–R2 potential divider. The op-amp is shown used in the unity-gain non-inverting mode, with an input impedance of 47k (=R3).

Note in the above two circuits that the external load must have an impedance of at least 1k0. If the external speaker has an impedance less than this, resistor R_x can

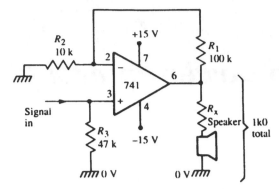

Figure 7.1 *Low-power amplifier using dual power supplies*

be connected as shown to raise the impedance to the 1k0 value, but inevitably reduces the amount of power reaching the actual speaker.

In practice, the available output current (and thus power) of an op-amp can easily be boosted via a transistor complementary emitter-follower network wired into its output; *Figure 7.3* shows such a circuit, using a single-ended supply. Here, the Q1–Q2 emitter follower has slight forward bias applied via D1 and D2 and is wired into the circuit's negative feedback loop to minimise cross-over distortion, etc. The circuit gives unity overall voltage gain and can supply output currents up to 350mA peak or 50mA r.m.s. into a load of 23R minimum, i.e. it can provide powers up to 280mW into such a load.

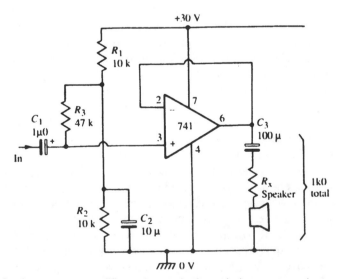

Figure 7.2 *Low-power amplifier using a single-ended power supply*

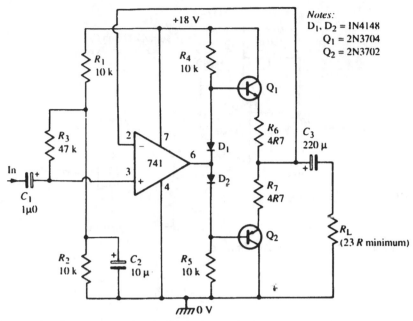

Figure 7.3 *Op-amp power amplifier using a single-ended supply*

Simple audio PA ICs

If output powers greater than a couple of hundred milliwatts are needed, the easiest solution is to use a dedicated audio power amplifier (PA) IC. A large range of such ICs are readily available, in both mono (single amplifier) and dual (two amplifiers) forms, and with maximum output power ratings ranging from about 325mW to 68W. *Figures 7.4* to *7.7* show practical application circuits using mono ICs with power ratings in the 325mW to 5W range.

The *Figure 7.4* circuit is based on the LM386, which is housed in an 8-pin DIL package and has an output power rating of 325mW; it can use DC supplies in the 4V to 15V range, consumes a quiescent current of 4mA, and is useful in many battery-powered applications. The diagram shows the IC used in the non-inverting mode; the voltage gain can be set via pins 1 and 8, and equals ×20 when these pins are open-circuit, or ×200 when they are ac-shorted via C4. Note that C1 is used to RF decouple the positive supply pin (pin-6), and R1–C3 is an optional Zobel network that ensures high frequency (HF) stability when feeding an inductive speaker load.

The *Figure 7.5* circuit is based on the LM388, which is as a modified version of the LM386 fitted into a 14-pin DIL package with an integral heat sink connected to pins 3–4–5 and 10–11–12. The IC can operate from 4V to 12V supplies and can feed 1.5W into an 8R0 load when using a 12V supply. The diagram shows the IC used in

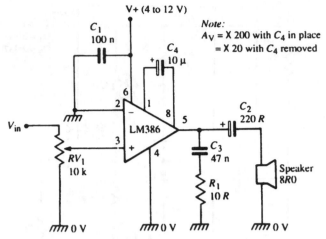

Figure 7.4   *An LM386 (325mW) amplifier*

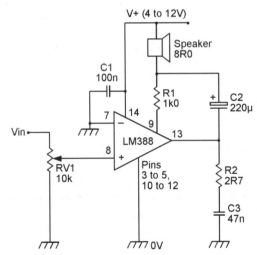

Figure 7.5   *A simple LM388 (1.5W) amplifier*

the minimum-components configuration, with the speaker connected to the positive supply line and the circuit's voltage gain set at ×20. R2–C3 is an optional Zobel network.

The *Figure 7.6* circuit is also based on the LM388, but in this case the speaker is wired between the IC's output and ground. The IC is (as in the case of the *Figure 7.5* circuit) used in the non-inverting voltage amplifying mode. The voltage gain can optionally be set via pins 2 and 6, and equals ×20 when these pins are open-circuit, or ×200 when they are ac shorted via C5.

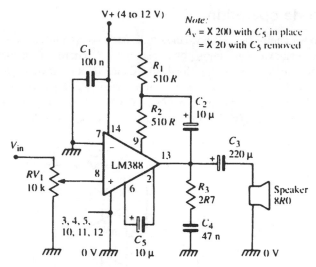

Figure 7.6 *An LM388 (1.5W) amplifier*

Finally, the *Figure 7.7* circuit can use either the LM380 or the LM384 IC, and can supply output powers up to a maximum of 5W. The LM380 is probably the best known of all audio PA ICs; it can use any supply in the 8V to 22V range, and can deliver 2W into an 8R0 load when operated with an 18V supply (but needs a good external heat sink to cope with this power level). It has ground-referenced differential input terminals, a fully protected short-circuit proof output, and gives a fixed voltage gain of ×50 (34 dB). The LM384 is simply an uprated version of the LM380, capable of operating at supply values up to 26V and delivering 5.5W into an external load. Both types of IC are housed in 14-pin DIL packages in which pins 3–4–5 and 10–11–12 are meant to be thermally coupled to an external heat sink.

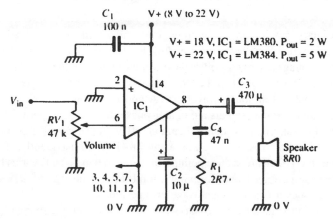

Figure 7.7 *An LM380/384 (2W or 5W) amplifier*

Bridge-mode operation

When an IC power amplifier is used in the single-ended output mode shown in *Figures 7.4* to *7.7*, the peak available output power equals V^2/R, where V is the peak available output voltage and R is the output load impedance. Note, however, that available output power can be increased by a factor of four by connecting a pair of amplifier ICs in the 'bridge' configuration shown in *Figure 7.8*, in which the peak available load power equals $(2V)^2/R$. This power increase can be explained as follows.

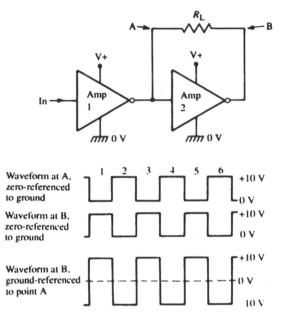

Figure 7.8   *A pair of amplifiers connected in the bridge mode give a peak output of $(2V)^2/R$ watts, i.e., four times the power of a single-ended circuit*

In an ordinary single-ended amplifier, one end of the output load is usually grounded; in *Figure 7.8*, however, both ends of load R_L are floating and driven in anti-phase, and the voltage across R_L equals the difference between the A and B values. The diagram shows the waveforms applied to the load via a 10V peak-to-peak squarewave input signal. Note that although waveforms A and B each have peak values of 10V relative to ground, the two signals are in anti-phase; thus, during period 1 point B is 10V positive to A and is thus seen as being at +10V, but in period 2 point B is 10V negative to point A and is thus seen as being at –10V. Consequently, if point A is regarded as a zero voltage reference point, it can be seen that point B varies from +10V to –10V between periods 1 and 2, giving a total voltage change of 20V across R_L. Similar changes occur in subsequent waveform periods.

Thus, the load in a 10V bridge-driven circuit sees a total voltage swing of 20V peak-to-peak, or twice the single-ended voltage value. Since doubling the drive voltage results in a doubling of drive current, and power is equal to the *V–I* product, the bridge-driven circuit thus produces four times more power output than a single-ended one.

Figures 7.9 and *7.10* show a couple of practical examples of bridge-driven audio power amplifiers. The *Figure 7.9* circuit uses a pair of LM388 ICs and can feed 4W into an 8R0 load when using a 12V supply. The *Figure 7.10* circuit can use a pair

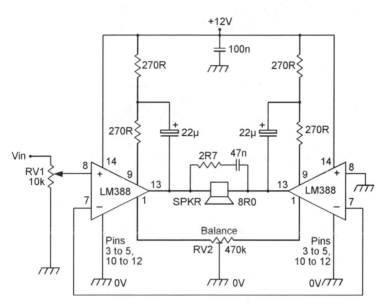

Figure 7.9 *LM388 bridge amplifier delivers 4W into an 8R0 load*

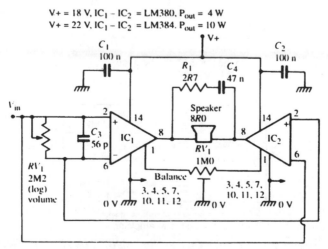

Figure 7.10 *4W or 10W bridge-configured amplifier*

LM380 or LM384 ICs, and can deliver up to 10W into an 8R0 load. Note in each of these circuits that the speaker is direct coupled to the outputs of the power amplifier ICs, eliminating the need for bulky and expensive electrolytic decoupling capacitors.

Dual IC applications

Audio power amplifier ICs are widely available in both mono and dual forms. Dual types are very handy for making stereo or bridge-driven amplifiers, and the LM2877

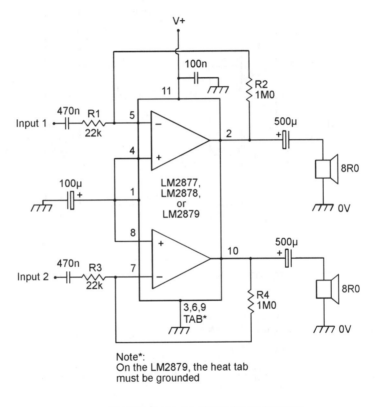

Note*:
On the LM2879, the heat tab
must be grounded

| | IC Type | | |
	LM2877	LM2878	LM2878
V+ (max)	24V	32V	32V
P_{out}/channel	4W	5W	8W
A_v	50	50	50
Z_{in}	22k	22k	22k

Typical performance of the
inverting sterio amplifier

Figure 7.11   *Simple stereo amplifier using the LM2877, LM2878, or LM2879 dual amplifier ICs*

dual 4W, the LM2878 dual 5W, and the LM2879 dual 8W are amongst the best known of these types. These three devices differ mainly in their voltage/power ratings and packaging styles; the three ICs are each housed in an 11-pin single-in-line (SIL) package with an attached heat sink, but the LM2879 package is particularly robust.

The LM2877/78/79 range of ICs is very easy to use: the input stages of each half are meant to be dc biased to half-supply volts, and a bias generator is built into each IC for this purpose. *Figure 7.11* shows how to wire each IC as a simple stereo amplifier that is powered from a single-ended supply and drives an 8R0 loudspeaker load. Each amplifier is wired in the inverting mode, is biased by connecting its non-inverting input pin to the IC's bias terminal (pin-1), and has its closed-loop voltage gain set at ×50 by the ratio of R2/R1 or R4/R3. The table shows the typical performance of this circuit.

If desired, the output power of each IC half can be boosted to a higher value via a couple of external power transistors; *Figure 7.12* shows how the available output power of one-half of the LM2878 can be boosted to 15W in this way. This remarkably simple circuit generates a typical THD of only 0.05 percent at 10W output; at very low power levels Q1 and Q2 are inoperative and power is fed to the speaker via R2; at higher power levels Q1 and Q2 act as a normal complementary emitter-follower and provide most of the power drive to the speaker. R2 and the base-emitter junctions of Q1–Q2 are effectively wired into the circuit's negative feedback loop, thus minimising signal cross-over distortion.

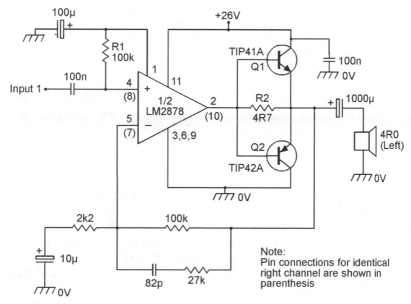

Figure 7.12 *One channel of a 15W per channel stereo amplifier using a single-ended supply*

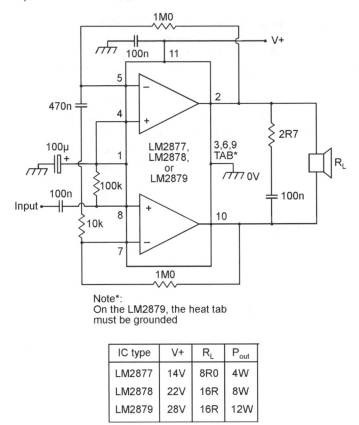

Figure 7.13 *Bridge amplifier using dual ICs*

IC type	V+	R_L	P_{out}
LM2877	14V	8R0	4W
LM2878	22V	16R	8W
LM2879	28V	16R	12W

Finally, *Figure 7.13* shows how the two halves of an LM2877, LM2878, or LM2879 can be used to make a bridge-configured mono amplifier which can feed relatively high power levels to a direct-coupled speaker load.

ICs for cars

Several manufacturers produce audio power amplifier ICs for specific use in in-car entertainment systems. Two of the best-known devices of this type are the identical LM383 and TDA2003 8W devices, which are each housed in a 5-pin TO-220 plastic package with integral heat sink. At a car's normal 'running' supply voltage of 14.4V each IC can deliver 5.5W into a 4R0 load or 8.6W into a 2R0 load; each IC can in fact operate with any 5V to 20V supply voltage and can supply peak output currents of 3.5A.

The LM383 (TDA2003) is an easy device to use. *Figure 7.14* shows it wired as a 5.5W in-car amplifier, with the IC wired in the non-inverting mode with its closed-

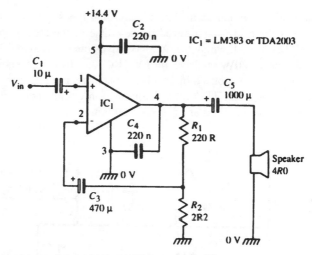

Figure 7.14 *LM383 (TDA2003) 5.5W in-car amplifier*

loop voltage gain set at ×100 via R1–R2. C2 and C4 ensure the IC's HF stability; C4 must be direct wired between pins 3 and 4.

If desired, a pair of LM383 or TDA2003 ICs can be wired in the bridge configuration, to act as a 16W in-car mono amplifier. *Figure 7.15* shows the circuit connections; note that pre-set pot RV1 is used to balance the quiescent output voltages of the two ICs, to minimise the circuit's quiescent operating current.

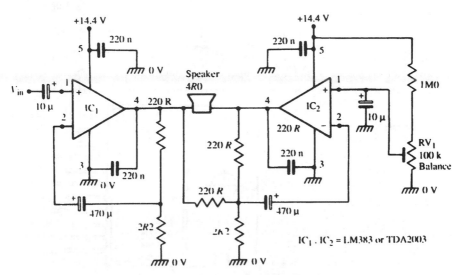

Figure 7.15 *LM383 (TDA2003) 16W in-car amplifier*

Some of the available in-car audio PA ICs are dual types, which can be used in either the stereo or the bridge-configured modes. One of the most popular type of 'dual' IC is the TDA2005M, which has its two halves internally wired in the bridge mode, to provide up to 20W of drive into a 2R0 load from the vehicle's 14.4V (nominal) supply. The IC is housed in an 11-pin package, as shown in *Figure 7.16*, which also shows a practical TDA2005M application circuit; note that all of this circuit's capacitors must be rated at 25 volts minimum.

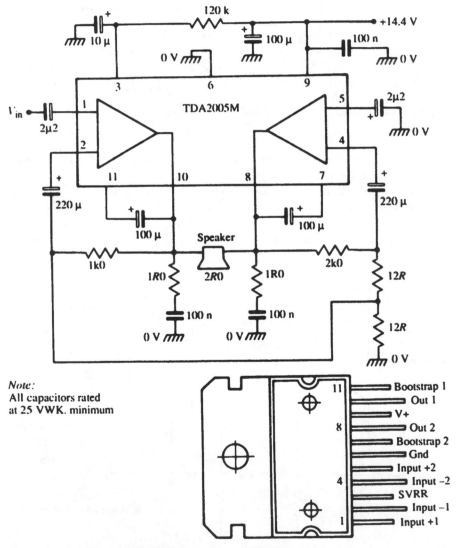

Figure 7.16   *TDA2005M 20W in-car power booster*

Hi-Fi amplifiers

Several manufacturers produce high-performance mono audio power amplifier ICs that are suitable for direct use in Hi-Fi systems with per-channel output power ratings in the range 8W to 68W. To conclude this chapter, *Figures 7.17* to *7.20* show four practical application circuits for ICs of this type.

The *Figure 7.17* circuit has an output power rating of 8W, and is designed around a TDA2006 IC. This popular high-quality device typically generates less than 0.1 percent distortion when feeding 8W into a 4R0 speaker, and is housed in a 5-pin TO-220 package with an electrically insulated heat tab which can be

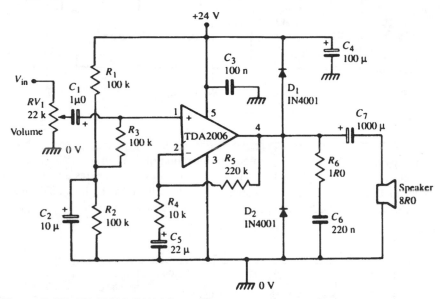

Figure 7.17 *TDA2006 8W hi-fi amplifier*

bolted directly to an external heat sink without need of an insulation washer. In the diagram, the IC's non-inverting input pin is biased at half-supply volts via R3 and the R1–R2 potential divider, and the voltage gain is set at $\times 22$ via R5/R4. D1 and D2 protect the IC's output against damage from the speaker's back e.m.f.s, and R6–C6 form a Zobel network.

The *Figure 7.18* circuit is almost identical to the above, but has an output power rating of 15W and is designed around a TDA2030 IC. This very popular IC can be regarded as an uprated version of the TDA2006, and is housed in the same 5-pin TO-220 package with insulated heat tab. It can operate with single-ended supplies of up to 36V; when used with a 28V supply it gives a guaranteed output of 12W into 4R0, or 8W into 8R0. Typical THD is 0.05 percent at 1kHz at 7W.

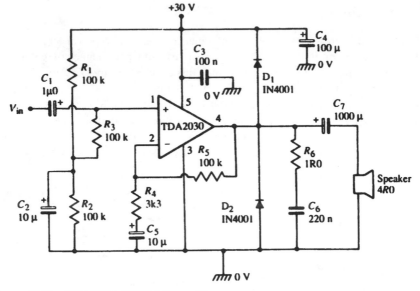

Figure 7.18   *TDA2030 15W hi-fi amplifier*

The *Figure 7.19* circuit has an output power rating of 20W into a 4R0 load, and is designed around an LM1875 IC. This high-quality device generates a mere 0.015 percent THD at 20W output, can operate at supply voltages as high as 60V, and is housed in a 5-pin TO-220 package with an insulated heat tab.

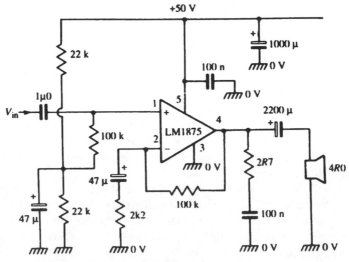

Figure 7.19   *LM1875 20W hi-fi amplifier*

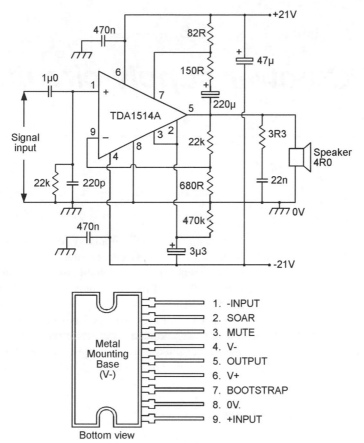

Figure 7.20 *TDA1514A 40W hi-fi amplifier*

Finally, the *Figure 7.20* circuit has an output power rating of 40W into a 4R0 load, and is designed around a TDA1514A. This is a very high performance IC that typically generates a mere 0.0032 percent distortion at 16W output into a 4R0 load. The IC is housed in a 9-pin package (shown in the diagram) with an integral heat sink that is internally connected to the IC's negative supply pin (pin-4); the heat sink must be insulated from ground in all split-supply applications, such as that illustrated.

8

DC power supply circuits

This chapter deals with low- to medium-power DC supply systems and circuits of the type that are either powered from existing DC energy sources, or derive their power from the normal 50–60Hz AC supply line via a suitable isolating transformer (systems that generate DC power directly from the AC supply line are described in Chapter 9). The chapter is divided into four main sections that, in sequence, deal with conventional ways of deriving DC supplies from AC power lines, with DC voltage regulator circuits, with low-power voltage converter circuits that can (for example) generate a higher-voltage or reversed-polarity supply from an existing DC power source, and finally, with modern switched-mode DC voltage regulators.

AC/DC converter basics

There are two basic ways of deriving a stable low- to medium-power DC supply from a normal AC power line, and these are shown in basic form in *Figures 8.1* and *8.2*. The 'conventional' way (*Figure 8.1*) is to use a step-down transformer and a rectifier and storage capacitor to generate (with an overall efficiency of about 85 percent) an unregulated DC supply that is electrically insulated from the AC supply,

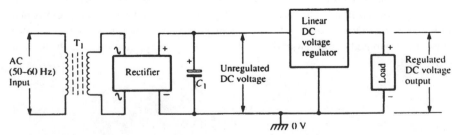

Figure 8.1 *Conventional regulated DC power supply; typical conversion efficiency is 45 percent*

and then to stabilise the DC output voltage via a linear regulator circuit, which typically has an efficiency slightly better than 50 percent, thus giving an overall power conversion efficiency of about 45 percent. Such systems are simple and reliable, generate zero RFI, and give excellent ripple reduction and voltage regulation, but are rather bulky and heavy, because (a) the transformer operates at the 50–60Hz supply line frequency, and (b) a heat sink is needed to dissipate the power wasted by the low conversion efficiency. Systems of this type are used in most radio and audio equipment, and in all RFI-sensitive instruments.

A variant of the above system uses a switched-mode voltage regulator in place of the linear component; such systems operate with an overall efficiency of about 80 percent. They generate significant RFI, but are widely used in desk-top calculators and computers and in other compact equipment that is not particularly RFI sensitive. Some practical systems of this type are described at the end of this chapter.

The alternative AC/DC conversion system uses the switched-mode technique shown in *Figure 8.2*. Here, the AC power line voltage is directly converted to DC via a rectifier and storage capacitor, and this DC is used to power a series-connected 20kHz switched-mode (variable M/S-ratio) voltage regulator and isolating trans-former, which has its output converted back to DC via another rectifier and storage

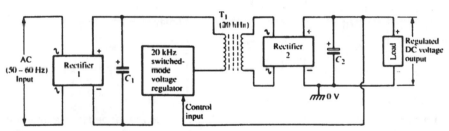

Figure 8.2 *Switched-mode regulated DC power supply; typical conversion efficiency is 80 percent*

capacitor; part of the DC output is fed back to the switched-mode voltage regulator, to complete a control loop. This system has a typical efficiency of about 80 percent and (because its heat-sink requirements are minimal and its isolating transformer works at 20kHz) can be very compact, but it generates substantial RFI and gives poorer ripple rejection and voltage regulation than the conventional system. Systems of this type are now rarely used.

Basic power supply circuits

Most modern electronic equipment uses the 'conventional' type of power supply in which the AC/DC converter consists of a transformer that converts the AC line voltage into an electrically insulated and more useful AC value, and a rectifier–filter combination that converts this into smooth DC of the desired voltage value, and this is the only type considered in this chapter. *Figures 8.3* to *8.6* show the four most widely used basic power supply circuits. The *Figure 8.3* design provides a single-ended DC supply from a single-ended transformer and bridge–rectifier combination, and gives a performance virtually identical to that of the *Figure 8.4* centre-tapped transformer circuit. The *Figure 8.5* and *8.6* circuits each provide split or dual DC supplies with

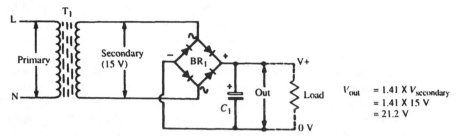

Figure 8.3   *Basic single-ended power supply using a single-ended transformer and bridge rectifier*

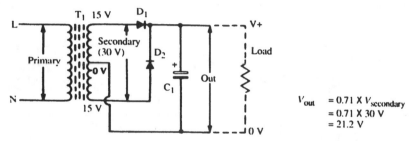

Figure 8.4   *Basic single-ended power supply using a centre-tapped transformer and two rectifiers*

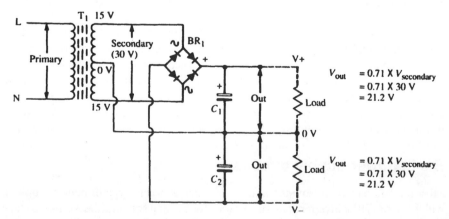

Figure 8.5   *Basic split or dual power supply using a centre-tapped transformer and bridge rectifier*

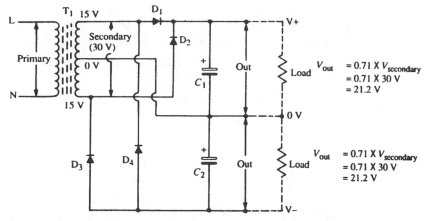

Figure 8.6 *Basic split or dual power supply using a centre-tapped transformer and individual rectifiers*

nearly identical performances. The rules for designing these four circuits are quite simple, as follows.

Transformer–rectifier selection

The three most important parameters of a transformer are its secondary voltage, its power rating, and its regulation factor. The secondary voltage is always quoted in r.m.s. terms at full rated power load, and the power load is quoted in terms of VA or watts. Thus, a 15V 20VA transformer gives a secondary voltage of 15V r.m.s. when its output is loaded by 20W. When the load is removed (reduced to zero) the secondary voltage rises by an amount implied by the *regulation factor*. Thus, the output of a 15V transformer with a 10 percent regulation factor (a typical value) rises to 16.5V when the output is unloaded.

Note that the transformer's r.m.s. output voltage is *not* the same as the DC output voltage of the complete full-wave rectified power supply which, as shown in *Figure 8.7*, is in fact 1.41 times greater than that of a single-ended transformer, or 0.71 times that of a centre-tapped transformer (ignoring rectifier losses). Thus, a single-ended 15V r.m.s transformer with 10 percent regulation gives an output of about 21V at full rated load (just under 1A at 20VA rating) and 23.1V at zero load. When rectifier losses are taken into account the output voltages are slightly lower than shown in the graph. In the two-rectifier circuits of *Figures 8.4* and *8.6* the losses are about 600mV, and in the bridge circuits of *Figures 8.3* and *8.5* they are about 1.2V. For maximum safety, the rectifiers should have current ratings at least equal to the DC output currents.

Thus, to select a transformer for a particular task, first decide the DC output voltage and current that is needed, to establish the transformer's minimum VA rating, then simply consult the graph of *Figure 8.7* to find the transformer secondary r.m.s. voltage that corresponds to the required DC voltage.

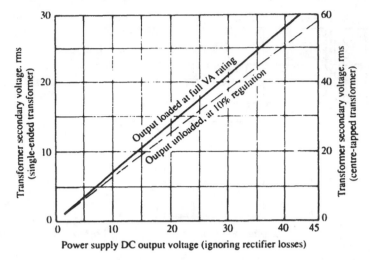

Figure 8.7 *Transformer selection chart. To use, decide on the required loaded DC output voltage (say 21V), then read across to find the corresponding transformer secondary voltage (15V single ended or 30V centre tapped)*

The filter capacitor

The purpose of the filter capacitor is to convert the rectifier's output into a smooth DC voltage; its two most important parameters are its working voltage, which must be greater than the off-load output value of the power supply, and its capacitance value, which determines the amount of ripple that will appear on the DC output when current is drawn from the circuit.

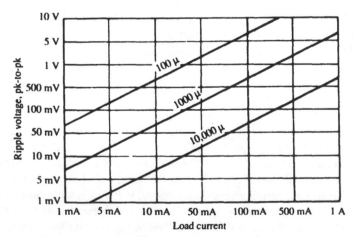

Figure 8.8 *Filter capacitor selection chart, relating capacitor size to ripple voltage and load current in a full-wave rectified 50–60Hz powered circuit*

As a rule of thumb, in a full-wave rectified power supply operating from a 50–60Hz power line, an output load current of 100mA will cause a ripple waveform of about 700mV peak-to-peak to be developed on a 1000µF filter capacitor, the amount of ripple being directly proportional to the load current and inversely proportional to the capacitance value, as shown in the design guide of *Figure 8.8*. In most practical applications, the ripple should be kept below 1.5V peak-to-peak under full load conditions. If very low ripple is needed, the basic power supply can be used to feed a 3-terminal voltage regulator IC, which can easily reduce the ripple by a factor of 60dB or so at low cost.

Zener-based voltage regulators

Practical voltage regulators vary from simple Zener diode circuits designed to provide load currents up to only a few milliamps, to fixed- or variable-voltage high-current circuits designed around dedicated 3-terminal voltage regulator ICs. Circuits of all these types are shown in the next few sections of this chapter.

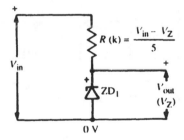

Figure 8.9 *This basic Zener 'reference' circuit is biased at about 5mA*

Figure 8.9 shows how a Zener diode can be used to generate a fixed reference voltage by passing a current of about 5mA through it from the supply line via limiting resistor R. In practice, the output reference voltage is not greatly influenced by sensible variations in the diode current value, and these may be caused by variations in the values of R or the supply voltage, or by drawing current from the output of the circuit. Consequently, this basic circuit can be made to function as a simple voltage regulator, generating output load currents up to a few tens of milliamps, by merely selecting the R value as shown in *Figure 8.10*.

Here, the value of R is selected so that it passes the maximum desired output current plus 5mA; consequently, when the specified maximum output load current is being drawn the Zener passes only 5mA, but when zero load current is being drawn it passes all of the R current, and the Zener dissipates maximum power; the power rating of the Zener must not be exceeded under this 'no load' condition.

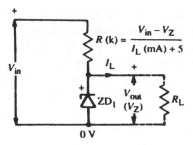

Figure 8.10   *This basic Zener 'regulator' circuit can supply load currents of a few tens of milliamps*

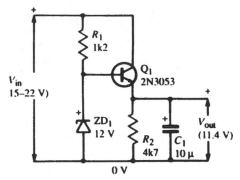

Figure 8.11   *This series-pass Zener-based regulator circuit gives an output of 11.4V and can supply load currents up to about 100mA*

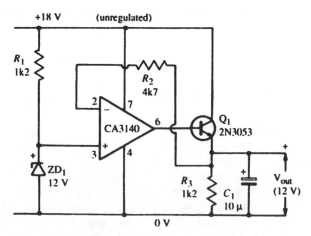

Figure 8.12   *This op-amp based regulator gives an output of 12V at load currents up to 100mA and gives excellent regulation*

The available output current of a Zener regulator can easily be increased by wiring a current-boosting voltage follower into its output, as shown in the series-pass voltage regulator circuits of *Figures 8.11* and *8.12*. In *Figure 8.11*, Q1 acts as the voltage following current booster, and gives an output that is about 600mV below the Zener value; this circuit gives reasonably good regulation. In *Figure 8.12*, Q1 and the CA3140 op-amp form a precision current-boosting voltage follower that gives an output equal to the Zener value under all load conditions; this circuit gives excellent voltage regulation. Note that the output load current of each of these circuits is limited to about 100mA by the power rating of Q1; higher currents can be obtained by replacing Q1 with a power Darlington transistor.

Fixed 3-terminal regulator circuits

Fixed-voltage regulator design has been greatly simplified in recent years by the introduction of 3-terminal regulator ICs such as the '78xxx' series of positive regulators and the '79xxx' series of negative regulators, which incorporate features such as built-in fold-back current limiting and thermal protection, etc. These ICs are available with a variety of current and output voltage ratings, as indicated by the 'xxx' suffix; current ratings are indicated by the first part of the suffix (L = 100mA, blank = 1A, S = 2A), and the voltage ratings by the last two parts of the suffix (standard values are 5V, 12V, 15V and 24V). Thus, a 7805 device gives a 5V positive output at a 1A rating, and a 79L15 device gives a 15V negative output at a 100mA rating.

Three-terminal regulators are very easy to use, as shown in *Figures 8.13* to *8.15*, which show positive, negative and dual regulator circuits respectively. The ICs shown are 12V types with 1A ratings, but the basic circuits are valid for all other voltage values, provided that the unregulated input is at least 3V greater than the desired output voltage. Note that a 270nF or greater disc (ceramic) capacitor must be wired close to the IC's input terminal, and a 10μF or greater electrolytic is wired

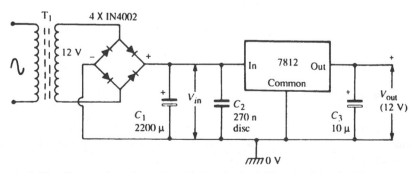

Figure 8.13 *Connections for using a 3-terminal positive regulator, in this case a 12V 1A '78' type*

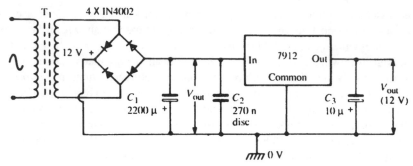

Figure 8.14 *Connections for using a 3-terminal negative regulator, in this case a 12V 1A '79' type*

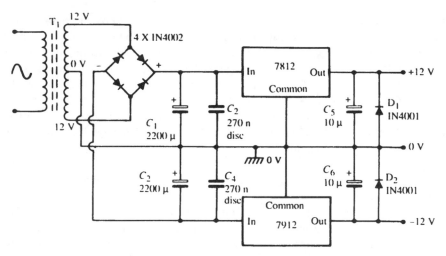

Figure 8.15 *Complete circuit of a 12V 1A dual power supply using 3-terminal regulator ICs*

across the output. The regulator ICs typically give about 60dB of ripple rejection, so 1V of input ripple appears as a mere 1mV of ripple on the regulated output.

Voltage variation

The output voltage of a 3-terminal regulator IC is actually referenced to the ICs 'common' terminal, which is normally (but not necessarily) grounded. Most regulator ICs draw quiescent currents of only a few milliamps, which flow to ground via this common terminal, and the IC's regulated output voltage can thus be raised above the designed value by biasing the common terminal with a suitable voltage, making it easy to obtain odd-ball output voltage values from these 'fixed voltage' regulators. *Figures 8.16* to *8.18* show three ways of achieving this.

In *Figure 8.16* the bias voltage is obtained by passing the IC's quiescent current (typically about 8mA) to ground via RV1. This design is adequate for many applications, although the output voltage shifts slightly with changes in quiescent current. The effects of such changes can be minimised by using the *Figure 8.17*

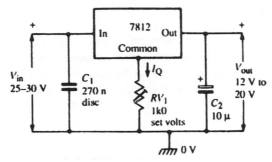

Figure 8.16 *Very simple method of varying the output voltage of a 3-terminal regulator*

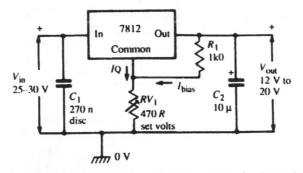

Figure 8.17 *An improved method of varying the output of a 3-terminal regulator*

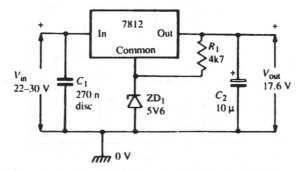

Figure 8.18 *The output voltage of a 3-terminal regulator can be increased by a fixed amount by wiring a suitable Zener diode in series with the common terminal*

design, in which the RV1 bias voltage is determined by the sum of the quiescent current and the bias current set by R1 (12mA in this example). If a fixed output with a value other than the designed voltage is required, it can be obtained by wiring a Zener diode in series with the common terminal as shown in *Figure 8.18*, the output voltage then being equal to the sum of the Zener and regulator voltages.

Current boosting

The output current capability of a 3-terminal regulator can be increased by using the circuit of *Figure 8.19*, in which current boosting can be obtained via by-pass transistor Q1. Note that R1 is wired in series with the regulator IC; at low currents insufficient voltage is developed across R1 to turn Q1 on, so all the load current is provided by the

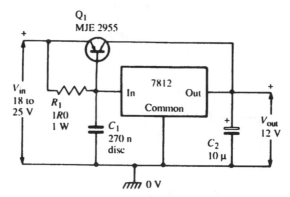

Figure 8.19 *The output current capacity of a 3-terminal regulator can be boosted via an external transistor. This circuit can supply 5A at a regulated 12V*

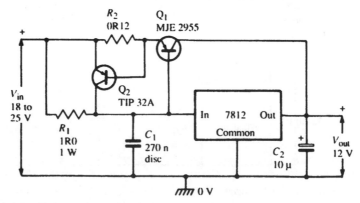

Figure 8.20 *This version of the 5A regulator has overload protection provided via Q2*

IC. At currents of 600mA or greater sufficient voltage (600mV) is developed across R1 to turn Q1 on, so Q1 provides all current in excess of 600mA.

Figure 8.20 shows how the above circuit can be modified to provide the bypass transistor with overload current limiting via 0R12 current-sensing resistor R2 and turn off transistor Q2, which automatically limit the output current to about 5A.

Variable 3-terminal regulator circuits

The 78xxx and 79xxx range of 3-terminal regulator ICs are designed for use in fixed-value output voltage applications, although their outputs can in fact be varied over limited ranges. If the reader needs regulated output voltages that are variable over very wide ranges, they can be obtained by using the 317K or 338K 3-terminal 'variable' regulator ICs.

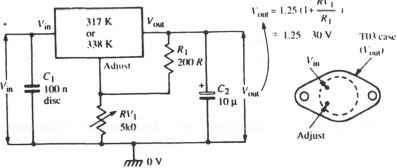

Parameter	317 K	338 K
Input voltage range	4–40 V	4–40 V
Output voltage range	1.25–37 V	1.25–32 V
Output current rating	1.5 A	5 A
Line regulation	0.02%	0.02%
Load regulation	0.1%	0.1%
Ripple rejection	65 dB	60 dB

$$V_{out} = 1.25 \left(1 + \frac{RV_1}{R_1} \right)$$

$$= 1.25 - 30 \text{ V}$$

Figure 8.21 *Outline, basic data and application circuit of the 317K and 338K variable-voltage 3-terminal regulators*

Figure 8.21 shows the outline, basic data and the basic variable regulator circuit applicable to these two devices, which each have built-in fold-back current limiting and thermal protection and are housed in TO3 steel packages. The major difference between the devices is that the 317K has a 1.5A current rating compared to the 5A rating of the 338K. Major features of both devices are that their output terminals are always 1.25V above their 'adjust' terminals, and their quiescent or adjust-terminal currents are a mere 50µA or so.

Thus, in the *Figure 8.21* circuit, the 1.25V difference between the 'adjust' and output terminals makes several milliamps flow to ground via R1, thus causing a variable voltage to be developed across RV1 and applied to the 'adjust' terminal. In practice, the output of this circuit can be varied from 1.25V to 33V via RV1, provided that the unregulated input voltage is at least 3V greater than the output. Alternative voltage ranges can be obtained by using other values of R1 and/or RV1, but for best stability the R1 current should be at least 3.5mA.

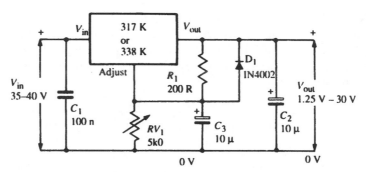

Figure 8.22 *This version of the variable-voltage regulator has 80dB of ripple rejection*

The basic *Figure 8.21* circuit can be usefully modified in a number of ways; its ripple rejection factor, for example, is about 65dB, but this can be increased to 80dB by wiring a 10μF bypass capacitor across RV1, as shown in *Figure 8.22*, together with a protection diode that stops the capacitor discharging into the IC if its output is short circuited.

Figure 8.23 shows a further modification of the *Figure 8.22* circuit; here, the regulator's transient output impedance is reduced by increasing the C2 value to

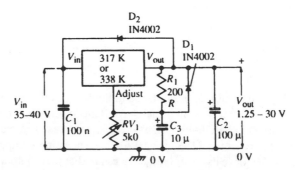

Figure 8.23 *This version of the regulator has 80dB ripple rejection, a low-impedance transient response, and full input and output short-circuit protection*

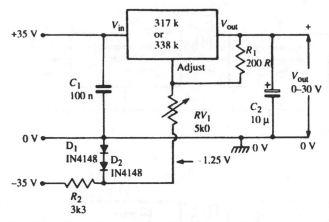

Figure 8.24 *The output of this version of the regulator is fully variable from zero to 30V*

100µF and using diode D2 to protect the IC against damage from the stored energy of this capacitor if an input short occurs.

Finally, *Figure 8.24* shows how the circuit can be modified so that its output is variable all the way down to zero volts, rather than to the 1.25V of the earlier designs. This is achieved by using a 35V negative rail and a pair of series-connected diodes that clamp the low end of RV1 to minus 1.25V.

Voltage converter basics

Electronic voltage converters are circuits used to generate a higher-value supply from an existing low-voltage DC source, or to generate a negative DC supply from an existing positive DC voltage source, etc. Such circuits are fairly easy to design and build, using either readily available components or special-purpose ICs such as the ICL7660 voltage converter. A variety of low-power versions of such circuits are shown in the remainder of this chapter.

A DC voltage can easily be converted into one of greater value or reversed polarity by using the DC supply to power a free-running squarewave generator which has its output fed to a multi-section capacitor–diode voltage multiplier network, which thus provides the desired 'converted' output voltage. If a positive output voltage is needed, the multiplier must give a non-inverting action, as in *Figure 8.25(a)*, and if a negative output is required it must give an inverting action, as in *Figure 8.25(b)*.

Practical converters of this type can use a variety of types of multivibrator circuit (bipolar or FET transistor, CMOS or TTL IC, etc.) as their basic free-running squarewave generators; in all cases, however, the generator should operate at a frequency in the range 1kHz to 10kHz, so that the multiplier section can operate with good efficiency while using fairly low values of 'multiplying' capacitor.

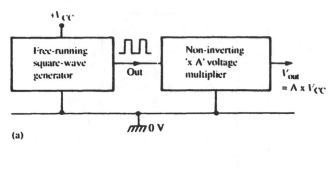

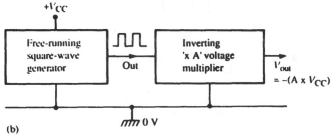

Figure 8.25 (a) *Voltage converter with positive output, and* (b) *voltage converter with negative output*

One of the easiest ways of making practical voltage converters of this type is to use type-555 'timer' ICs (which can supply fairly high output currents) as the free-running squarewave generators, and *Figures 8.26* to *8.29* show a selection of practical circuits of this type; in each case the 555 is wired as a free-running astable multivibrator and operates at about 3kHz (determined by the R1–R2–C2 values); C1 and C3 both help enhance circuit stability.

The *Figure 8.26* circuit acts as a DC voltage doubler, and generates a DC output voltage roughly double that of the 555's supply line via the C4–D1–C5–C2 capacitor-diode voltage-doubler network, which produces the $2 \times V_{cc}$ output voltage.

This '$2 \times V_{cc}$' is the approximate value of the unloaded output voltage; the precise value equals $2 \times V_{peak}$, minus ($V_{df1} + V_{df2}$), where V_{peak} is the peak output voltage of the squarewave generator and 'V_{df}' is the forward volt drop (about 600mV) of each 'multiplier' diode. The output voltage decreases when the output is loaded.

The *Figure 8.26* circuit can be used with any DC supply in the range 5V to 15V and can thus provide a 'voltage-doubled' output of 10V to 30V. Greater outputs can be obtained by adding more multiplier stages to the circuit. *Figure 8.27*, for example, shows how to make a DC voltage tripler, which can provide outputs in the range 15V to 45V, and *Figure 8.28* shows a DC voltage quadrupler, which gives outputs in the 20V to 60V range.

A particularly useful type of 555 converter is the DC negative-voltage generator, which produces an output voltage almost equal in amplitude but opposite in polarity

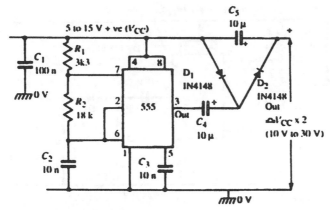

Figure 8.26 *DC voltage doubler circuit*

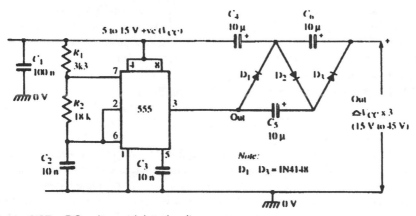

Figure 8.27 *DC voltage-tripler circuit*

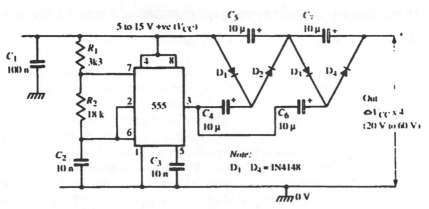

Figure 8.28 *DC voltage quadrupler circuit*

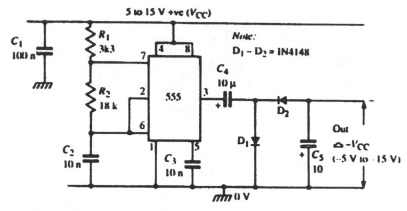

Figure 8.29 *DC negative-voltage generator*

to that of the IC supply line. This type of circuit can be used to provide a split-supply output for powering op-amps, etc., from a single-ended power supply. *Figure 8.29* shows an example of such a circuit, which operates at 3kHz and drives a voltage-doubler (C4–D1–C5–D2) output stage.

High-voltage generation

The 'voltage multiplier' method of generating increased values of output voltage is usually cost effective only when multiplier ratios of less than six are needed. In cases where very large step-up ratios are required (as, for example, when hundreds of volts must be generated via a 12V supply), it is often better to use the output of a low-voltage oscillator or squarewave generator to drive a step-up voltage transformer, which then provides the required high-value voltage (in AC form) on its secondary (output) winding; this AC voltage can easily be converted back to DC via a simple rectifier–filter network, if required. *Figures 8.30* to *8.32* show some practical low-power high-voltage generator circuits of these types.

The *Figure 8.30* circuit acts as a DC-to-DC converter which generates a 300V DC output from a 9V DC power supply. In this case Q1 and its associated circuitry act as a Hartley *L–C* oscillator, with the low-voltage primary winding of 9V–0–9V to 250V mains transformer T1 forming the '*L*' part of the oscillator, which is tuned via C2. The supply voltage is stepped up to about 350V peak at T1 secondary, and is half-wave rectified and smoothed via D1–C3. With no permanent load on C3, the capacitor can deliver a powerful but non-lethal 'belt'. With a permanent load on the output, the output falls to about 300V at a load current of a few milliamps.

The *Figure 8.31* circuit can be used to drive a neon lamp or generate a low-current high-value (up to a few hundred volts) DC voltage from a low-value (5V to 15V) DC supply. The 555 is wired as a 3kHz astable multivibrator that drives T1 via R3.

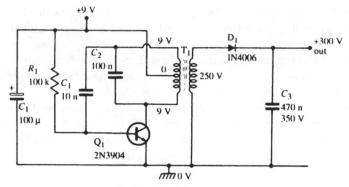

Figure 8.30 *9V to 300V DC-to-DC converter*

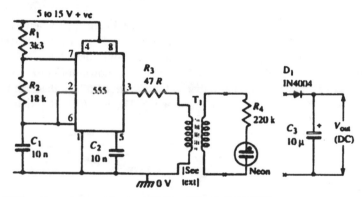

Figure 8.31 *Neon-lamp driver or 'high-voltage' generator*

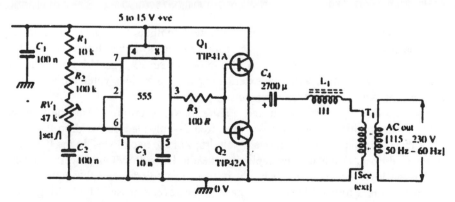

Figure 8.32 *DC-to-AC inverter*

T1 is a small audio transformer with a turns ratio sufficient to give the desired output voltage; for example, with a 10V supply and a 1:20 T1 turns ratio, the transformer will give an unloaded AC output of 200V peak; this AC voltage can easily be converted to DC via a half-wave rectifier and filter capacitor, as shown.

Finally, the *Figure 8.32* DC-to-AC inverter circuit produces an AC output at normal AC power-line frequency and voltage. The 555 is wired as a low-frequency (variable from 50Hz to 60Hz via RV1) astable that feeds its power-boosted (via Q1–Q2) output into the low-voltage 'input' of reverse-connected filament transformer T1, which has the desired 'step-up' turns ratio. C4 and L1 act as a filter that ensures that the power signal feeding into the transformer is essentially a sine wave.

The ICL7660

The ICL7660 is a dedicated voltage converter IC specially designed to generate an equal-value negative supply from a positive source, i.e. if powered from a +5V supply it generates a –5V output. It can be used with any +1.5V to 10V DC supply, and has a typical voltage conversion efficiency of 99.9 percent(!) when its output is unloaded; when the output is loaded it acts like a voltage source with a 70R output impedance, and can supply maximum currents of about 40mA.

The ICL7660 is housed in an 8-pin DIL package as shown in *Figure 8.33*. The IC actually operates in a way similar to that of the *Figure 8.26* 'oscillator and voltage-multiplier' circuit, but with far greater efficiency. The ICL7660 chip houses a very

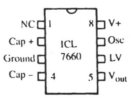

Figure 8.33   *Outline and pin notations of the ICL7660 voltage converter IC*

efficient squarewave generator that operates (without the use of external components) at about 10kHz and has an output that switches fully between the supply rail values; it also houses an ultra-efficient set of logic-driven multiplier 'diodes' that, when used with two external capacitors, enables voltage doubling to be achieved with near-perfect efficiency.

The reader may recall from the *Figure 8.26* description that the use of conventional multiplier diodes makes the circuit's unloaded output drop by about 1.2V, this being the sum of the forward volt drops of the two diodes. In the ICL7660 this volt drop is eliminated by replacing the diodes with MOS power switches, driven via logic networks in such a way that each 'diode' switch automatically closes when it is forward biased and opens when reverse biased, thus giving near-perfect operating efficiency.

The ICL7660 is an easy device to use, but note that none of its terminals must ever be connected to a voltage greater than V+ or less than GROUND. If the IC is to be used with supplies in the range 1.5V to 3.5V, the pin-6 'LV' terminal should be grounded; at supply values greater than 3.5V, pin-6 must be left open circuit. At supply values greater than 6.5V a protection diode must be wired in series with OUTPUT pin-5. *Figures 8.34* to *8.42* show a selection of practical application circuits in which these design rules are applied.

ICL7660 circuits

The most popular application of the ICL7660 is as a simple negative-voltage converter, and *Figures 8.34* to *8.36* show three basic circuits of this type; in each case, C1 and C2 are 'multiplier' capacitors and each have a value of 10μF. The *Figure 8.34* voltage converter is intended for use with 1.5V to 3.5V supplies, and requires the use of only two external components. The *Figure 8.35* circuit is similar, but is meant for use with supplies in the 3.5V to 6.5V range and thus has pin-6

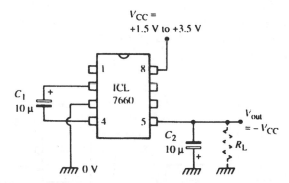

Figure 8.34 *Negative-voltage converter using 1.5V to 3.5V supply*

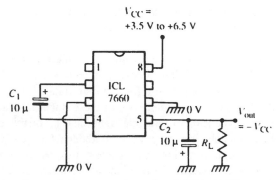

Figure 8.35 *Negative-voltage converter using 3.5V to 6.5V supply*

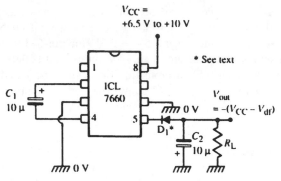

Figure 8.36   *Negative-voltage converter using 6.5V to 10V supply*

grounded. Finally, the *Figure 8.36* circuit is meant for use with supplies in the range 6.5V to 10V, and thus has diode D1 wired in series with output pin-5, to protect it against excessive reverse biasing from C2 when the power supplies are removed. The presence of this diode reduces the available output voltage by V_{df}, the forward volt drop of the diode; to keep this volt drop to minimum values, D1 should be a germanium or Schottky type.

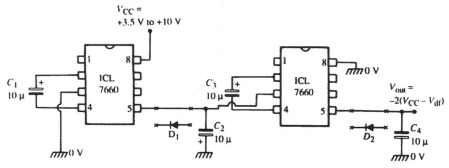

Figure 8.37   *Cascading devices for increased output voltage*

A useful feature of the ICL7660 is that numbers of these ICs (up to a maximum of ten) can be cascaded to give voltage conversion factors greater than unity. Thus, if three stages are cascaded, they give a final output voltage of $-3V_{cc}$, etc. *Figure 8.37* shows the connections for cascading two of these stages; any additional stages should be connected in the same way as the right-hand IC of this diagram.

In some applications the user may want to reduce the ICL7660's oscillator frequency; one way of doing this is to wire capacitor C_x between pins 7 and 8, as in *Figure 8.38*; *Figure 8.39* shows the relationship between the Cx and frequency values; thus, a C_x value of 100pF reduces the frequency by a factor of 10, from

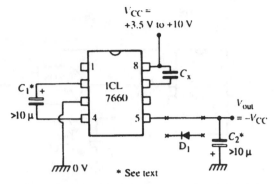

Figure 8.38 *Reducing oscillator frequency*

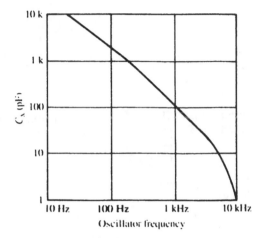

Figure 8.39 *C_x/oscillator frequency graph*

10kHz to 1kHz; to compensate for this 10:1 frequency reduction and maintain the circuit efficiency, the C1 and C2 values should be increased by a similar factor (to about 100µF each).

Another way of reducing the oscillator frequency is to use pin-7 to overdrive the oscillator via an external clock, as shown in *Figure 8.40*. The clock signal must be fed to pin-7 via a 1k0 series resistor (R1), and should switch fully between the two supply rail values; in the diagram, a CMOS gate is wired as an inverting buffer stage, to ensure such switching.

Another use of the ICL7660 IC is as a positive voltage multiplier, to give a positive output of almost double the original supply voltage value. *Figure 8.41* shows the circuit connections. The pin-2 oscillator output signal is used here to drive a conventional capacitor–diode voltage-doubler network, of the type used in *Figure 8.26*. Note that these two diodes reduce the available output voltage by an amount

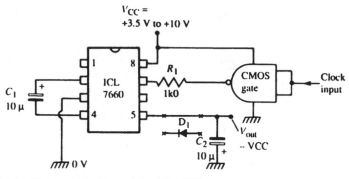

Figure 8.40   *External clocking of the ICL7660*

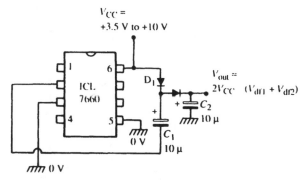

Figure 8.41   *Positive-voltage multiplier*

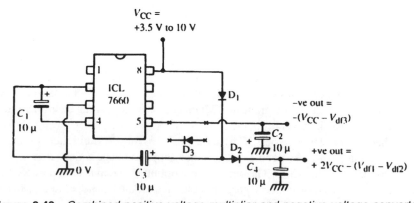

Figure 8.42   *Combined positive-voltage multiplier and negative voltage converter*

equal to their combined forward volt drops, so they should ideally be low-loss germanium or Schottky types.

Finally, to complete this look at the ICL7660, *Figure 8.42* shows how the circuits of *Figures 8.35/8.36* and *8.41* can be used to make a combined positive-voltage multiplier and negative-voltage converter that provide dual output voltage rails; each rail has an output impedance of about 100R.

Switched-mode voltage regulators

In the opening section of this chapter it was pointed out that a very efficient way of generating a stable DC supply is first to derive a crude DC supply from the AC power line via a conventional transformer and rectifier–capacitor network, and then to convert that crude DC into a well-regulated form via a switched-mode voltage regulator. *Figure 8.43* shows the basic circuit of this type of system. Here, C1's unregulated DC voltage is fed to the input of the switched-mode voltage regulator, which has its output smoothed by an *L–C* filter system and applied to a feedback loop that helps to regulate the final output voltage value.

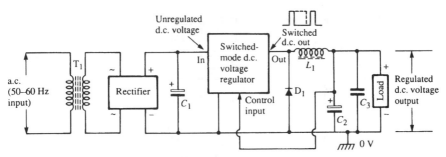

Figure 8.43 *Basic switched-mode regulated DC power supply circuit*

Figure 8.44 shows a more detailed view of the actual switched-mode voltage regulator circuitry. Here, the unregulated DC voltage is applied to a variable M/S-ratio (often called a 'pulse-width modulated', or PWM) generator, which typically operates at about 20kHz and drives series-pass transistor Q1, so that a powerful variable M/S-ratio waveform with a peak amplitude of V_{pk} is generated at the D1–L1 junction. D1 forms part of the $L_1–C_1$ filter network, which integrates this M/S waveform and gives a smooth DC output voltage of $V_{pk} \times M/(M + S)$. Self-regulation is achieved by feeding part of the DC output back to the generator's *Control* input terminal.

The *Figure 8.44* type of circuit is very efficient, since Q1 operates as a power switch that wastes little energy and thus generates little heat, and the L1–C1 filter values can be quite small, since the generator operates at a fairly high frequency. In practice,

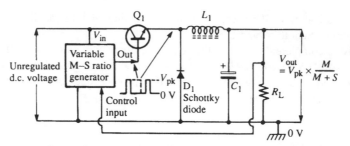

Figure 8.44   *Switched-mode voltage regulator circuit*

switched-mode regulator circuits are best built using dedicated ICs, and one of the most popular of such devices is the L296, which is shown in *Figure 8.45*.

The L296 is a sophisticated switched-mode voltage regulator capable of supplying a 5.1V to 40V output at 4A maximum, and houses a sawtooth oscillator, a 5.1V voltage reference and comparator and various logic and sensing circuits, all housed in a 15-lead plastic power package with an integral ground-connected metal

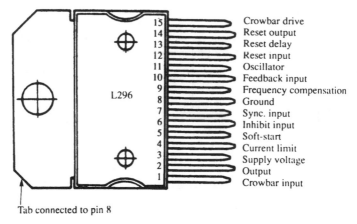

Tab connected to pin 8

Figure 8.45   *Outline and pin notations of the L296 switched-mode voltage regulator IC*

mounting tab. *Figure 8.46* shows the simplest (minimum component count) way of using the L296, as a 5.1V regulator with a 4A current rating; C2–R1 set the sawtooth oscillator frequency to 100kHz, C3–R2 set the feedback time constants, C4 gives a slow-start action (at initial power-up), and the regulated output is set at 5.1V (the internal reference value) by shorting pin-10 to the output. Note that D1 is a Schottky type, that C1 and C5 need high ripple current ratings, and that signal and power grounds must be made as shown.

Figure 8.47 shows the circuit modified to give a variable (5.1V to 15V) output by making the feedback connection via the RV1–R3 divider ($V_{out} = 5.1V \times [RV1+R3]/$

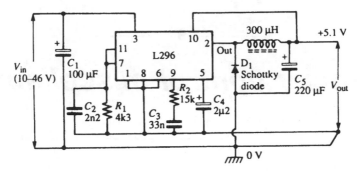

Figure 8.46 *Minimum-count 5.1V/4A switched-mode voltage regulator; note the* signal *and* power *ground paths used in this circuit*

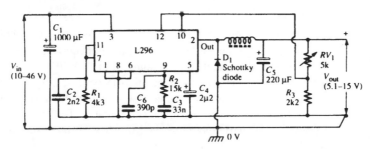

Figure 8.47 *Variable (5.1V to 15V) 4A switched-mode regulated DC voltage supply circuit*

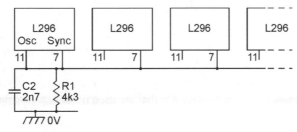

Figure 8.48 *Multiple L296 operations using a single oscillator (to avoid inter-modulation problems)*

R3) and by adding C6; pin-10 is shorted to pin-12, to give the circuit automatic start-up reset.

Finally, to complete this brief look at the L296, note in cases where several of these ICs are to be used in a single complex circuit (to generate several independent output voltages), that several L296 ICs can use the same oscillator (thus avoiding intermodulation problems) by using the system shown in *Figure 8.48*, where the oscillator of one L296 is used to drive the pin-7 'Sync' pins of the others, which have their pin-11 'Osc' terminals left open.

$$9$$

Power control miscellany

Many of the earlier chapters of this book have looked at circuits that can be used for controlling the electrical power that is derived from the conventional single-phase 50–60Hz AC power lines, either directly or via some form of AC/DC converter. No mention has yet been made of the various practical aspects of dealing with the AC power lines, or of the basic types of power lines that are in common use. This final chapter deals with both of these subjects, gives some practical advice on the use/ modification of domestic AC wiring systems, and finishes off by describing some early SCR-based AC power-control circuits.

AC power distribution basics

In all modern countries, national electrical power supplies are generated in extremely high-voltage (EHV) 3-phase AC form (usually 50–60Hz) and are then distributed nationwide to the consumers via a network of cables and step-down voltage transformers. The network takes the basic form shown in *Figure 9.1*, which also shows (in brackets) the specific voltages that are used in the UK distribution system.

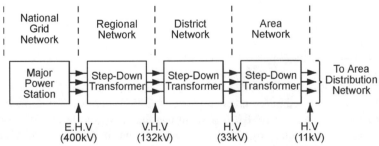

Figure 9.1 *A single distribution path of a national AC electrical power distribution system (UK voltages are shown in brackets)*

Thus, major power stations feed their 3-phase power into the *national* grid at 400kV, and individual *regional* consumer boards then tap into the grid via 400kV/132kV step-down transformers and distribute the 132kV 3-phase VHV (very-high-voltage) supplies to their own individual *district* consumer boards via VHV cable networks. The individual *district* consumer boards then tap into the VHV network via 132kV/33kV step-down transformers and distribute the resulting 33kV 3-phase HV (high-voltage) supplies to their own individual district *areas*, which in turn further reduce the voltage via 33kV/11kV step-down transformers before further processing the supply and distributing it to *individual consumers* such as industrial, commercial, and domestic units within those areas. Note that this system is designed to minimise current-generated distribution power losses caused by the resistance of the distribution network's supply cables; thus, an end consumer who draws 1A (11kW) from the local 11kV supply will impose a drain of only 27.5mA on the national grid, which may derive its power from a generating plant that is hundreds of miles away.

Figure 9.2 shows the basic form of a typical *area* AC electrical distribution path, in which HV supplies are distributed to various consumer groups. Thus, the 33kV supply that feeds the input of the area's 33kV/11kV step-down transformer may also be fed directly to heavy industrial units that have their own step-down transformer

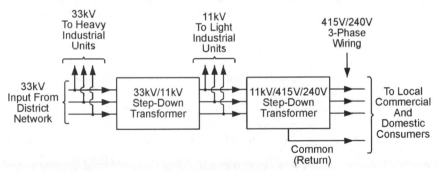

Figure 9.2 *A typical UK area AC power distribution path*

and power-distribution systems, and the 11kV output of the step-down transformer may be fed to light industrial units that have their own AC power processing systems. The 11kV 3-phase AC distribution system is also fed to local commercial and domestic consumers via local sub-stations, which employ a step-down transformer to provide a '415V/240V' 3-phase 4-wire output from the 11kV 3-phase 3-wire input.

Figure 9.3 shows basic details of the above-mentioned '415V/240V' 3-phase 4-wire AC power supply system. Important points to note here are that the AC voltages on the three supply lines are 120° out of phase with each other, that the r.m.s. voltage between each of the *phase* lines is 415V, and that the r.m.s. voltage

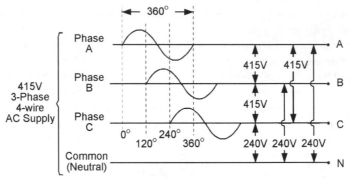

Figure 9.3   *Basic details of the '415V/240V' 3-phase 4-wire AC power supply system used in the UK*

between each *phase* line and the *common* (neutral) line is 240V. This type of system can thus be used for driving high-power 3-phase 415V electric motors by connecting the motor to all three *phase* lines, or can be used to supply normal 240V AC power to domestic or commercial premises (etc.) via the *common* line and any one of the three *phase* lines.

One other important feature of the *Figure 9.3* AC power supply system is that, because of the 120° phase shift that exists between the three supply lines, the sum of the three instantaneous voltages that appear between *common* and the three *phase* lines always equals zero. When, for example, the 'A'-line voltage is at its peak value of +340V, the 'B' and 'C' voltages are each at –170V, thus giving an overall sum of zero volts. Consequently, if identical loads are connected between each *phase* line and *common*, each *phase* line will supply an identical r.m.s current, but zero r.m.s current will flow in the *common* line. These simple facts form the basis of the normal 240V AC domestic power distribution system, which is shown in *Figure 9.4*.

In *Figure 9.4*, the area supplied by a local sub-station transformer is divided into three consumer zones or sectors, each of which consumes approximately one-third

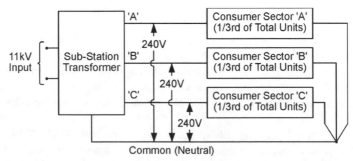

Figure 9.4   *Basic details of the 240V AC domestic power supply system used in the UK*

of the total electrical units absorbed by the area. Each of these sectors is connected to the 240V AC supply provided by one specific phase ('A', 'B', or 'C') of the 3-phase supply, and via the sub-station's *common* line. Thus, if the system is perfectly balanced the *common* line will always carry near-zero current, and if the system is severely disrupted (by the temporary removal of one complete sector from the system) the *common* line will carry a current no greater than that of the larger of the two remaining *phase* currents. Consequently, this basic 3-phase system allows any desired number of consumers to be supplied with a 240V AC single-phase 2-wire (plus 'earth') power supply, using only a 4-wire distribution system in which each wire requires a maximum current rating equal to one-third of the peak total current consumed by the system. This system is thus highly efficient in terms of power distribution costs.

Domestic AC wiring systems

Most houses are, as described above, provided with an ordinary single-phase AC power supply system. In the UK, this is actually a 50Hz 240V supply that is regulated to within 6 percent of 240V (i.e. within the limits 224V to 256V), but is often called a '230V' supply to harmonise with the nominal 230V 'norm' of other European Community countries. The supply is connected via a 3-wire *'phase, neutral* and *ground* (earth)' distribution network. A few large domestic premises (and many industrial units) are also supplied with a 3-phase 415V supply, for operating heavy-duty electric motors (for powering lifts, etc.).

In the UK, all 'modern' house wiring is designed to conform to safety regulations that were first laid down in 1947; these regulations have since been upgraded in fine detail, but are unchanged in essence. Thus, if you live in an old house that has not been rewired since 1947, it is probable that your old wiring has now reached a point where its rubber insulation has perished to the state where a total breakdown is imminent and is in danger of bursting into flames, and an urgent rewire – to modern standards – is probably required.

In all domestic UK buildings, electric power is fed into the house from the external 240V AC network and applied, via the building's electricity meter, to the individual household circuits via individual fuse units or, in modern systems, via a multi-fused *consumer* unit. The household circuits are, for convenience, classified as three distinct types: (a) power circuits (feeding power sockets, etc.), (b) lighting circuits and (c) accessory circuits (individual heavy-duty circuits feeding immersion heaters, cooker panels, etc).

Old-style (pre-1947) electrical systems use a *radial* system of power circuit wiring, as shown in *Figure 9.5*. Here, each individual power socket or small group of sockets is treated as an individual circuit and is connected to the AC power line via its own specific fuse and length of 3-core cable; often, the phase ('live') and neutral lines of each circuit are individually fused. Such systems are inefficient; they require excessive amounts of cable and many fuses.

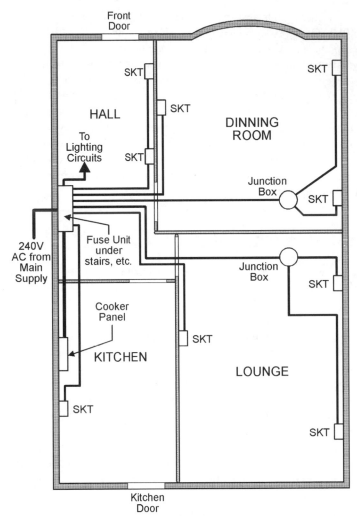

Figure 9.5   *Typical old (pre-1947) 'radial'-style ground-floor power wiring layout of a medium-sized UK house*

Modern (post-1947) UK systems of power wiring use a *ring* system of power-socket wiring. *Figure 9.6* shows a typical example of such a system, in which the ring can be connected directly to the fuse box (consumer unit) or can be connected via a junction box, as shown. Here, a 3-core power cable is run from the junction box to the first (hall) socket, then sequentially to all of the sockets on (say) the ground floor and finally back to the junction box, so that a 'ring' of cable is formed. The ring is treated as a single circuit and provided with a single master fuse, but each socket plug is individually fused. Most modern houses are provided with two

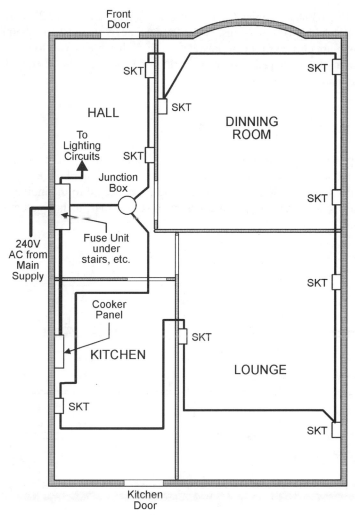

Figure 9.6 *Typical modern (post-1947) 'ring'-style ground-floor power wiring layout of a medium-sized UK house*

ring circuits, one to each floor. On the upper floor the ring may be connected to the fuse unit via a junction box and feeder cable.

The ring system is very efficient in the use of 3-core cable. Power is fed to each individual socket via both sides of the ring, which are thus effectively wired in parallel and share the socket current. This factor enables a relatively light gauge of ring cable to be used. In practice, 2.5mm^2 PVC sheathed cable is used for most ring circuits, and any desired number of standard 13A sockets can be fitted to such a ring, provided that it does not serve a floor area greater than 100m^2; such a ring must be fitted with a 30A (maximum) master fuse and can handle up to 7.2kW. The feeder cable to such a ring (if used) must be a 4.0mm^2 type. Note that most 2.5mm^2 cable has a maximum current rating of only 20A, so if such cable is used to make a multi-socket 'spur' that is tapped into the main ring, the spur must be fitted with its own 20A (maximum) master fuse.

Modern power circuit wiring

Rewiring the power circuitry of an average house to modern standards is a technically fairly simple task, but involves much dirty and hard work. Normally, the power ring is laid below the floorboards, and is connected to the individual power sockets by using either 'loop' techniques, or 'box-and-spur' techniques, as shown in *Figure 9.7*.

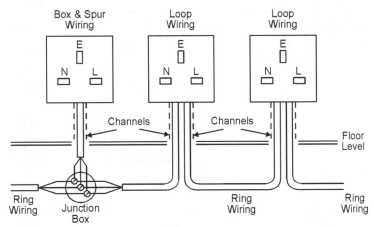

Figure 9.7   *Alternative ways of connecting a power socket to the main AC power ring*

Before starting such a rewire, plan very carefully the precise layout of each basic wiring ring, and the desired number and positions of all of its power sockets. Flush-type sockets are fitted into special metal boxes, which must (usually) be set into a wall or plaster, in which a channel must also be cut to accommodate the socket's power wiring, which is fed upwards from the sub-floor power ring. The socket can be loop wired into the ring by (1) making a loop in the ring's wiring, (2) feeding the loop up to the socket, then (3) cutting the loop so that two effective lengths of 3-core cable are available; finally (4) the tips of the six 'cores' can be bared and wired to the socket as shown in *Figure 9.8*. Note that this loop technique requires two lengths of cable to be buried in the socket's wall channel; the alternative 'box-and-spur' technique (*Figure 9.7*) requires only one length of cable to be so buried, the other end of the cable being connected into the ring via a sub-floor junction box.

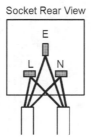

Figure 9.8 *Basic way of connecting 'loop' wiring to a power socket*

Modern lighting circuit wiring

The basic AC lighting circuit of a domestic system is deceptively simple, as shown in *Figure 9.9*. Note in particular that the light switch is used to connect the lamp to the live (L) side of the supply, thereby ensuring that an electric shock cannot be received from the lamp (when changing a bulb) when the switch is open ('off'). Also note that all lamp-plus-switch circuits are wired in parallel, across the supply lines, in 'radial' form.

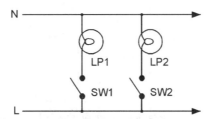

Figure 9.9 *Basic domestic lighting circuit ('earth' wiring not shown)*

The physical implementation of the *Figure 9.9* circuit is not simple. *Figure 9.10* shows the old-style (pre-1947) method of implementation, in which neutral (N) goes to one side of each lamp and the live line is looped from one switch to the next. Note that this method requires three wires plus earth (E) to be run down to each switch. In practice, the old-style wiremen often ignored the 'polarity' rules and switched the neutral line, so that the lamp socket remained permanently (and dangerously) 'live'.

Figure 9.11 shows the modern wiring method, which requires only two wires (plus earth) to be run down to each switch. The system makes extensive use of 4-terminal ceiling roses or junction boxes. In practice, $1mm^2$ 3-core 'twin and earth' (T & E) cable is used for wiring most domestic lighting circuits.

Figure 9.12 shows – in simplified form – the basic modern ground-floor lighting layout of a medium-sized house. The power feed is taken from the main fuse box and runs, 'radial' style, to all of the roses of the ground-floor lights. Light switches

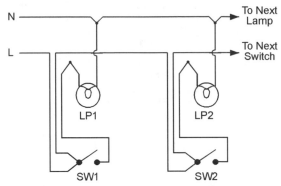

Figure 9.10   *Old-style implementation of the lighting circuit (earth wiring not shown), with three wires (plus earth) running down to each switch*

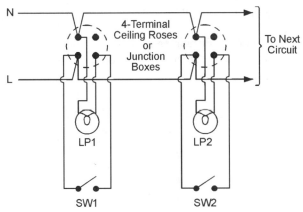

Figure 9.11   *Modern-style implementation of the lighting circuit (earth wiring not shown), with two wires (plus earth) running down to each switch*

are wired to the roses via additional cable. Once again, the system looks deceptively simple. In reality, the wiring to the roses runs above the ceiling; thus, the ground-floor's wiring must be laid by partially raising the carpets and floorboards on the first floor, and the wiring to the wall-mounted switches must be made by cutting suitable channels in the plaster, between the ceiling and the switches. Also, it is usual to apply multi-way switching (rather than switching from only a single point) to hall and landing lights, and this can be a very time-consuming task.

Figure 9.13 shows the circuit and implementation of 2-way light switching. In practice, 3-core 'twin and earth' cable is used to connect SW1 to the ceiling rose, and 4-core '3-core and earth' is used to interconnect SW1 and SW2, which are changeover switches. A lot of channelling work may be required to implement this system.

Figure 9.14 shows the electrical circuit involved in 3-way light switching, such as is often used in a hallway, in which the light can be controlled from either end

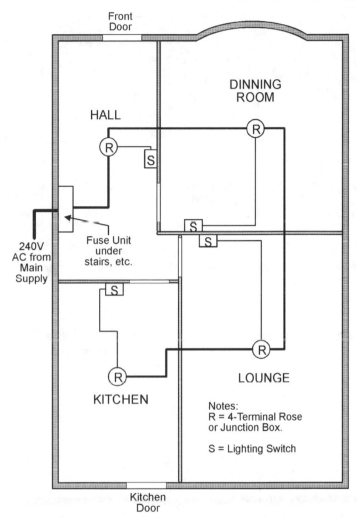

Figure 9.12 *Basic modern ground-floor lighting circuit of a medium-sized house*

of the hall or from an upper floor. This circuit uses two changeover switches and one standard 'intermediate' switch (available from most electrical retailers). This circuit is very time consuming to install and requires a lot of channelling.

Accessory circuit wiring

'Accessory' wiring is used for connecting fixed-wired (non-plugged) appliances to the AC power supply, and comes in two basic types, being for either less-than-13A *light-duty* use, or for greater-than-3kW (=12.5A at 240V) *heavy-duty* use.

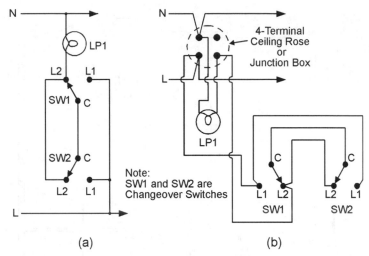

(a)   (b)

Figure 9.13 *Circuit* (a) *and implementation* (b) *of modern 2-way light switching (Earth wiring not shown)*

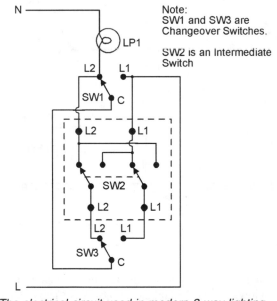

Figure 9.14 *The electrical circuit used in modern 3-way lighting*

Light-duty accessories include items such as electric door bells that are AC powered via step-down isolating transformers, electric clocks, small (less-than-3kW) fixed electric heaters and fans, hot-air hand dryers, air extractors, air-conditioning units, and dehumidifiers, etc. Such units may be connected to the fused consumer unit via individual wiring spurs, or can be connected to one of the main ring circuits. In the latter case, the accessory must be connected to the ring via a fused connection unit (often called a 'fused spur unit') and must be controlled via a double-pole switch, which may be incorporated in the accessory or the connection unit, or may be connected in series with the wiring.

Figure 9.15 shows three alternative ways of wiring fixed light-duty accessories to an existing ring circuit. In *(a)* the accessory incorporates a 2-pole switch and is connected to the ring via a simple fused connector unit and a junction box. In *(b)* the unswitched accessory is connected to the ring via a fused connector unit that incorporates a 2-pole switch via a junction box. In *(c)* the unswitched accessory is connected to a switched fused connection unit that is loop wired into the ring circuit.

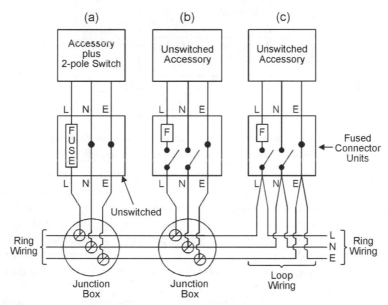

Figure 9.15 *Three alternative ways of wiring fixed light-duty accessories into an existing ring circuit*

Heavy-duty accessories include items such as electric cookers, water heaters, and storage heaters, etc., which must each be connected to the consumer unit via an individual wiring circuit and a 2-pole switch that is placed close to the accessory, in the manner shown in *Figure 9.16*. The circuit does not have to be individually fused, since overload protection will be incorporated in the part of the consumer unit that feeds the circuit.

Practical rewiring hints

Electrical rewiring is a technically simple but physically arduous, dirty and time-consuming task that can save the keen do-it-yourself enthusiast a huge amount of

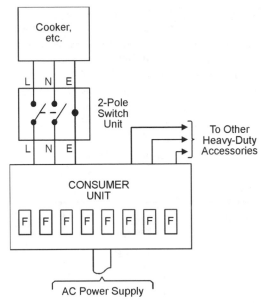

Figure 9.16   *Basic way of wiring a heavy-duty accessory to the AC power supply*

money. To gain access to underfloor wiring areas, you will usually have to move lots of furniture, lift large areas of carpet, and raise several floorboards in each room. You will have to crawl about beneath the floorboards, and may have to drill holes through joists and beams to facilitate wiring runs. If you are adding new flush-fitting power points or lighting switches to a house, plaster (and sometimes bricks) will have to be channelled to accept new cable runs and must be repaired afterwards. Essential tools for these jobs include two 2.5" bolster chisels, a lump hammer, and a good power drill.

Before starting your first rewire, buy or borrow a couple of good DIY books on the subject, and study them carefully. When you are ready, plan your new wiring layout with great care, giving lots of thought to the positioning of new power sockets, light switches, and ceiling roses, etc. In many houses, most of the original switch and power-socket wiring and floor-to-floor feeder cabling, etc., is fed via metal conduit (tubing), and your rewiring task may be greatly simplified by reusing this conduit, as explained in your DIY books.

Figures 9.17 to *9.19* show some useful rewiring aids. The *Figure 9.17* and *9.18* circuits can be used for tracing old wiring and conduit, and the *Figure 9.19* circuit acts as a 'marker beacon' that can be used to indicate the loft or underfloor break-through positions of pilot holes drilled through ceilings when installing new feeder cable runs or when repositioning ceiling roses, etc.

The *Figure 9.17* wiring tracer works on the magnetic-field detection principle and is used to trace 'live' wiring that – ideally – is actually passing current to a load. The wiring is located by sweeping across its suspected path with an ordinary telephone

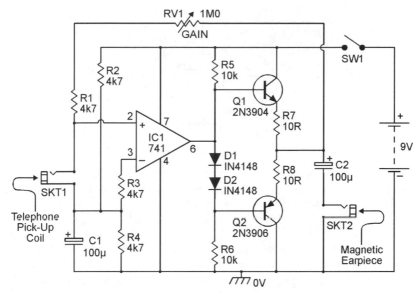

Figure 9.17 *This wiring tracer works on the magnetic-field detection principle*

pick-up coil until its 50Hz magnetic field induces a signal in the pick-up coil. This signal is amplified by op-amp IC1 and power boosted by the Q1–Q2 (etc.) amplifier, to finally produce a distinct 'hum' sound in the magnetic earpiece.

The *Figure 9.18* circuit acts as a simple BFO (beat-frequency oscillator) metal detector that is used in conjunction with a hand-held AM (long-wave or medium-wave) radio, and can be used to locate heavy-duty wiring or metal plumbing. The circuit is simply that of an *L–C* oscillator in which the inductance of L1 (and thus the frequency of oscillation) is influenced when L1 comes near most common types of metal. In use, the unit and the small radio are switched on and the radio is tuned until a

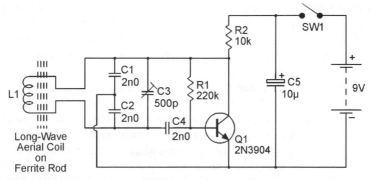

Figure 9.18 *Circuit of a simple BFO-type wiring and metal-pipe seeker; the unit is used in conjunction with a hand-held AM radio*

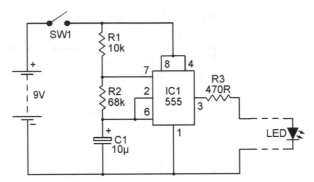

Figure 9.19 *This 'marker beacon' circuit can be used to indicate the positions at which pilot holes break through into dark places*

low 'beat' note is heard from its speaker; this 'beat' note should change when the unit's 'search head' (L1) is placed near metal. Hidden wiring (etc.) can then be located and its path traced by sweeping across its suspected route with the search head.

Finally, the *Figure 9.19* 'marker beacon' circuit consists of a simple low-frequency '555' astable multivibrator that pulses the LED on and off at a slow rate; the LED should be soldered to the end of a fairly stiff length of 2-core flex, so that it can easily be passed through small drill (pilot) holes, etc.

Early AC power control electronics

To conclude this book, the remaining parts of this final chapter look at early SCR-based electronic AC power control circuits of the types that are still sometimes found in industrial AC power control systems. Note that the actual term 'electronics' did not come into general use until the early 1940s, when virtually all electronic amplifier and control systems were designed around thermionic vacuum tubes ('valves'), and the most widely used type of electronic power switch was a fragile gas-filled thermionic tube called a thyratron. The first primitive transistor, a delicate point-contact type, was not invented until 1948, but it was not until 1952 that the first really big step into modern semiconductor electronics was made, when the first robust 'junction' type of transistor was invented by William Shockley, at the Bell Laboratories in the USA.

Shockley's basic junction transistor spawned a whole breed of brand new types of electronic amplifying and control devices, and in 1957 led to the development of the first commercially viable SCR (*S*ilicon-*C*ontrolled *R*ectifier), the first of a whole family of switching devices that were given the generic name 'thyristor', because they were transistor based and had thyratron-like characteristics. This family includes the triac, which acts like a bidirectional version of the SCR. Thyristors are inherently rugged and cost-effective devices, and by the early 1960s had made the fragile old thyratron obsolete.

The SCR is a simple four-layer pnpn silicon semiconductor device that can be simulated by a complementary pair of junction transistors and two resistors. *Figures 9.20* to *9.24* show diagrams that are relevant to the subject. Thus, *Figure 9.20* shows the symbol and basic three-layer construction of an npn junction transistor, *Figure 9.21* shows similar details of a pnp transistor, and *Figure 9.22* shows the symbol and

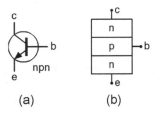

(a)　　　　　　　(b)

Figure 9.20 *Symbol* (a) *and basic 'layer' construction* (b) *of an npn junction transistor*

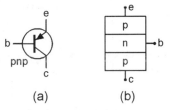

(a)　　　　　　　(b)

Figure 9.21 *Symbol* (a) *and basic 'layer' construction* (b) *of a pnp junction transistor*

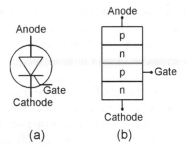

(a)　　　　　　　(b)

Figure 9.22 *Symbol* (a) *and basic 'layer' construction* (b) *of a pnpn SCR*

basic four-layer pnpn construction of the SCR. *Figure 9.23(a)* shows how the SCR's basic pnpn construction can be theoretically simulated by suitably interconnecting the layers of one pnp and one npn transistor, and *Figure 9.23(b)* shows how the SCR can actually be physically simulated by interconnecting two such transistors and two resistors in the form of a regenerative switch, in which Q1's collector current flows

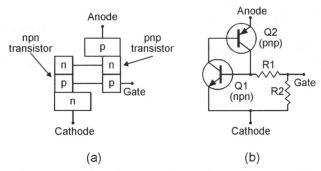

(a)   (b)

Figure 9.23   *Basic 2-transistor 'layer' analogue* (a) *and physical analogue* (b) *of the SCR*

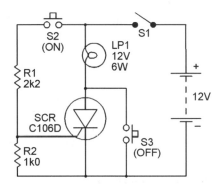

Figure 9.24   *Simple SCR DC on–off demonstration unit*

in Q2's base, and Q2's collector current flows in Q1's base. Finally, *Figure 9.24* shows a basic test circuit (using a readily available low-cost SCR) that can be used to demonstrate the SCR's basic DC characteristics, which can be understood by referring to *Figures 9.23(b)* and *9.24*, as follows.

(1) When power is first applied to the SCR by closing S1 in *Figure 9.24* the SCR is 'blocked' and acts (between anode and cathode) like an open-circuit switch. From *Figure 9.23(b)* it can be seen that this action occurs because Q1's base is shorted to its cathode via R1–R2, thus cutting off Q1 and depriving Q2 of base-current drive, hence cutting off Q2 and depriving Q1 of base-current drive.

(2) The SCR can be turned on by briefly closing S2 in *Figure 9.24*, thus applying gate current to the SCR via R1. This gate current causes the SCR to switch on regeneratively and act, between anode and cathode, like a forward-biased silicon rectifier with a saturation potential of about 1V, thus driving lamp LP1 fully on. This action can be understood with the aid of *Figure 9.23(b)*, from which it can be seen that the externally applied gate current feeds into the base of Q1, thus driving Q1 on and making it feed base current into Q2, which in turn

is driven on and starts feeding current directly into Q1's base, thus initiating a regenerative self-latching switching action that rapidly drives both transistors to saturation, to give an overall anode-to-cathode saturation voltage equal to the sum of Q2's saturation voltage (about 400mV) and Q1's base-emitter volt drop (about 600mV).

(3) Once the SCR has completed its regenerative switching action (the process takes only a microsecond or so, and only occurs when the anode current exceeds a 'minimum holding current' value of a few milliamps) the gate loses control, and the SCR self-latches into the 'on' state. Only a brief pulse of gate current is thus needed to activate the SCR, which exhibits very high gate-to-anode current gain. All of these points are fairly self-evident from *Figure 9.23(b)*, from which it can be seen that the gate-to-anode current gain is equal to the product of the Q1 and Q2 current gains. Typically, an SCR such as the C106D needs a maximum gate current of only 0.2mA at 1V, but has maximum anode ratings of 3.2A and 400V.

(4) Once the SCR has self-latched into the on state it can only be turned off again by briefly reducing its anode current below the device's 'minimum holding current' value; in *Figure 9.24* this can be achieved by briefly closing S3 or opening S1. Note that, in AC control applications, turn-off occurs automatically near the end of each positive AC half-cycle as the anode current falls below the critical 'minimum holding' value. In the *Figure 9.23(b)* 'analogue' circuit, the transistor current gains fall off as the anode current is reduced, and unlatching occurs when their combined loop gain falls below unity. Most practical SCRs have actual 'minimum holding' current values of a few milliamps.

Basic SCR AC power control circuits

Between the late 1950s and early 1970s the unidirectional SCR reigned supreme in AC power control applications, but towards the end of that period it began to take second place to the bidirectional triac in all but AC/DC converter and very high-power applications. Consequently, SCRs are now found mainly in either old AC power control circuits, or in some modern AC/DC converter or very high-power AC control circuits.

In most fully variable AC power control applications, SCRs are used in the 'phase-triggered' mode already described in Chapter 1, in which the SCR can be triggered on at any desired point during each AC half-cycle. There are several basic ways of using an SCR in such applications, and the most important of these are illustrated in *Figures 9.28* to *9.32*, which show the basic circuits used with DC loads. To understand fully the actions of these circuits, however, it is first necessary to note some fine points about the basic AC power supply, as shown in *Figure 9.25*, and about simple half-wave and full-wave rectifier circuits, as shown in *Figures 9.26* and *9.27*, as follows.

Figure 9.25 shows some important basic details about a resistively loaded 240V 50Hz AC supply. Note first that the term '240V' refers to the r.m.s. (root mean

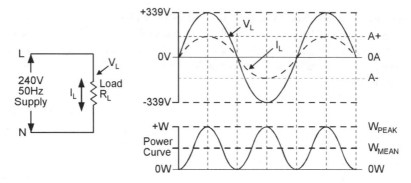

Figure 9.25 *Basic details of a resistively loaded 240V 50Hz supply*

square) or DC-equivalent power-generating value of the supply, which – when referenced to the 'N' terminal – swings alternately above and below the zero-volts reference by equal amounts, reaching a peak of 240V $\times$ 1.414 = 339V in each direction. The supply is a pure AC type, and thus has zero DC component. The load current (I_L) varies in direct proportion and polarity with that of the instantaneous supply voltage (V_L), causing the instantaneous load power to rise from zero at the start of each half-cycle, to a peak at the mid-point of each half-cycle, and to fall to zero again at the end of each half-cycle, thus producing two power peaks in each complete input cycle and generating a power curve with a fundamental frequency double that of the AC supply, i.e. 100Hz with a 50Hz supply.

Figure 9.26 shows basic details of a resistively loaded 240V 50Hz positive-output half-wave rectifier circuit. This circuit chops off the negative halves of the AC waveform, and the remaining voltage (V_L) consists of a DC component (V_{DC}) on

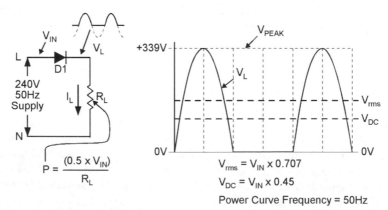

Figure 9.26 *Basic details of a resistively loaded 240V 50Hz positive output half-wave rectifier circuit*

which a large AC ripple component (V_{RIPPLE}) is superimposed; the r.m.s 'equivalent DC power-generating' value (V_{rms}) of this output is equal to the sum of these two components, and equals $V_{IN} \times 0.707$, where V_{IN} is the r.m.s. value of the original AC input voltage (240V). The power (in watts) actually absorbed by the resistive load is determined by V_{rms} (rather than V_{DC}), and – since $W = V^2/R$, and $V_{rms} = V_{IN} \times 0.707$ – equals $(0.5 \times V_{IN})/R_L$, and thus equals exactly half of the 'full' power that would be absorbed by the load if placed directly across V_{IN}. Note that the output power curve of the half-wave rectified circuit has a fundamental frequency equal to that of the AC supply, i.e. 50Hz in this case.

Figure 9.27 shows basic details of a resistively loaded 240V 50Hz positive-output full-wave bridge rectifier circuit. This circuit effectively converts negative input half-cycles into positive ones, leaving the positive half-cycles unchanged. The resulting output voltage (V_L) consists of a large DC component (V_{DC}) equal to $V_{IN} \times 0.9$, on which a 100Hz ripple component is superimposed. The V_{rms} value of this output equals the sum of these two components, and (ignoring inevitable circuit losses) exactly equals the original AC input voltage value, V_{IN}. The power absorbed by the load thus equals V_{IN}/R_L, and the output power curve has a fundamental frequency double that of the AC supply, i.e. 100Hz in this case.

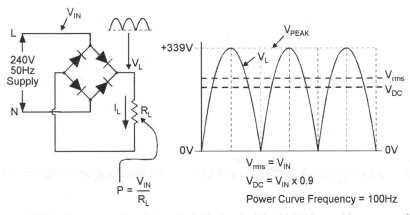

Figure 9.27 *Basic details of a resistively loaded 240V 50Hz positive output full-wave bridge rectifier circuit*

With all of the above data in mind, look now at the basic 'phase-triggered' SCR variable power control circuits shown in *Figures 9.28* to *9.32*. The basic controlled half-wave circuit of *Figure 9.28* acts like a controlled half-wave rectifier, and enables the load power to be varied from zero to half of the full-wave maximum value, but generates a large DC component at maximum output and generates a power-curve frequency of 50Hz that produces an annoying 'flicker' when applied to most filament lamps.

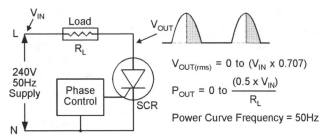

Figure 9.28   *Basic SCR-controlled half-wave circuit*

The *Figure 9.29* design is a variant of the above, with the SCR applying controlled positive half-cycles to the load, and rectifier D1 applying fixed (uncontrolled) negative half-cycles to the load, enabling the load power to be varied from 50 percent to 100 percent of the full-wave maximum value. The circuit generates a large DC component and an unbalanced 100Hz power curve over most of its range. The circuit's value can be increased by fitting switch SW1 in the position shown, enabling it to act like the *Figure 9.28* design when SW1 is open, thus covering the full 'zero-to-maximum' power driving range.

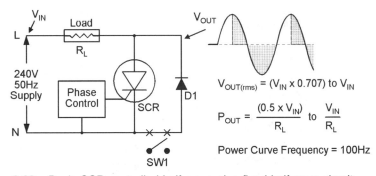

Figure 9.29   *Basic SCR-controlled half-wave plus fixed half-wave circuit*

Figure 9.30 shows a pair of SCRs used as a full-wave power controller. SCR1 controls the load's positive half-cycles, and SCR2 controls the negative half-cycles. If the two phase controllers are perfectly ganged, the output takes the form of a perfectly symmetrical AC waveform with a zero DC-component and a balanced fundamental power-curve frequency of 100Hz (perfect ganging can be achieved by using a single controller that drives the gates of both SCRs via a pair of pulse-transformer windings). The circuit's load power is fully variable from zero to V_{IN}/R_L.

Figure 9.31 shows a simple variant of the full-wave power controller, in which positive half-cycles are controlled via SCR_1 and D_2, and negative half-cycles via SCR2 and D1, and in which the two SCR gates are wired in parallel and driven via a single-phase controller. The circuit's performance is identical to that of *Figure 9.30*.

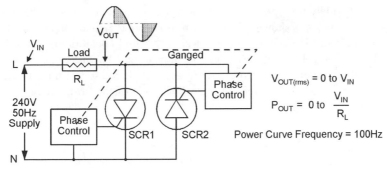

Figure 9.30 *Basic SCR-controlled full-wave circuit*

$V_{OUT(rms)} = 0$ to V_{IN}

$P_{OUT} = 0$ to $\dfrac{V_{IN}}{R_L}$

Power Curve Frequency = 100Hz

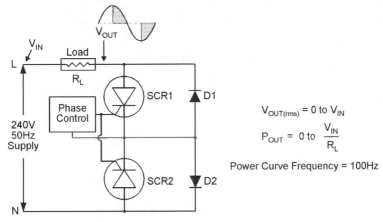

Figure 9.31 *Alternative SCR-controlled full-wave circuit*

$V_{OUT(rms)} = 0$ to V_{IN}

$P_{OUT} = 0$ to $\dfrac{V_{IN}}{R_L}$

Power Curve Frequency = 100Hz

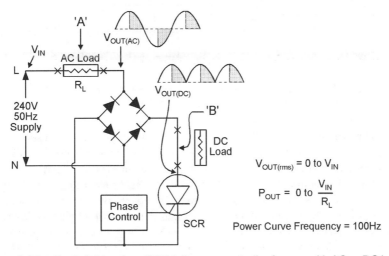

Figure 9.32 *Basic bridge-type SCR full-wave controller for use with AC or DC loads*

$V_{OUT(rms)} = 0$ to V_{IN}

$P_{OUT} = 0$ to $\dfrac{V_{IN}}{R_L}$

Power Curve Frequency = 100Hz

The most widely used present-day SCR power controller circuit is the full-wave 'bridge' type, which is shown in basic form in *Figure 9.32* and can be used to drive an AC load (placed in position 'A', with a short-circuit in position 'B') or a DC load (placed in position 'B', with a short-circuit in position 'A') with equal ease. Output power is fully variable from zero to V_{IN}/R_L via a single-phase control network (such as a UJT pulse generator driven from the 'DC' side of the bridge network).

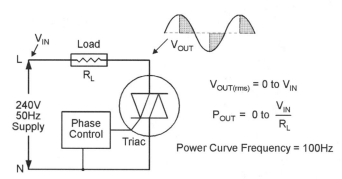

Figure 9.33 *Basic triac AC power control circuit*

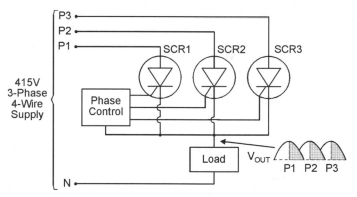

Figure 9.34 *Basic 3-phase half-wave controller using SCRs*

In reality, the *Figure 9.32* type of circuit is – nowadays – usually used only with DC loads, since AC loads are more efficiently driven by simple triac circuits of the basic type shown in *Figure 9.33*, many practical examples of which are shown in Chapter 4 of this volume. Finally, note that SCRs can be used to give variable half-wave control of 3-phase loads by using one SCR in each 'phase' line, using the basic connections shown in *Figure 9.34*, with the phase control of all three SCRs suitably 'ganged' and synchronised.

Index

Individual types of integrated circuits (ICs) mentioned in this Manual are listed in a separate section at the end of the Index.